# WHY MEN MAKE WAR

Richard E. Evans

Verity Books
New York

www.WhyMenMakeWar.com

Name: Richard E. Evans, 2023, author

Library of Congress Cataloging-in-Publication Data
Library of Congress Control Number: Pending

Title: Why Men Make War

Description: First Edition. New York: Verity Books

ISBN 978-1-7361733-2-9 (hardcover)
ISBN 978-1-7361733-3-6 (paperback)
ISBN 978-1-736173-3-1-2 (ePub)
ISBN 978-17361733-5-0 (Kindle)

SOC018000   SCI027000   POL047000

Subjects: Male and female behavior regarding violence and war.
Weapons and kinds of war. Policies to promote the advancement of
women in society and government. To reduce the number of wars,
women need to have as much political power as men.

Text and cover design by Bob Belinoff and Doug Van Pelt
Printed in the United States of America

2 4 6 8 9 7 5 3

# Three Battles

*(continued from cover)*

Still some distance away, a tall dark-skin stops and gathers the group around him, making sounds the hunters have never heard before. A few minutes later, about half the dark-skin men turn back toward the trees. The rest of the group holds its place, muttering among themselves and glancing at the curious Neanderthals, some of whom return to cutting off chunks of meat. The rest are wary, but not particularly concerned.

Suddenly, the dark-skins yell in unison and charge the Neanderthals, hurling spears as they run. The Neanderthals had never seen that before. They typically made their kills by thrusting their crude spears, or sometimes, by throwing from a few yards away, but not from what we would see as 30 to 40 yards away. Still, the Neanderthals stand their ground, unafraid of these lighter-built newcomers.

But some of the spears find their mark, and shouts of pain are heard. Just at that point, flying spears puncture the back of several Neanderthals. Too late, they realize the dark-skins who left the invading group have come up behind them. The out-numbered Neanderthals jab and thrust, but the dark-skins are faster and more nimble. They side-step the Neanderthal spears and ram or throw their own spears deep into Neanderthal flesh. The air is filled with screams of pain. In a few minutes, all the Neanderthals lie dead or dying. The dark-skins saunter back to their shelter with horse meat on their shoulders. Some elicit laughter from the group by mocking the Neanderthal cries of pain. Females and children rise up in joy to celebrate their return: food for everyone! That night, grunts and moans accompany a different kind of thrusting.

**Second battle**

Two eighteenth-century sailing ships, each armed with 40 cannons, ply the Atlantic Ocean off the North American coast. At some point, when still miles apart, crew members on both ships see the other's flag, indicating a country designated by a person they've never seen as "enemy."

No one on either ship has ever harmed anyone on the other ship. In fact, no one has ever met anyone on the other ship. Some people on both ships have families and children they love and care for – persons who would suffer if the men did not return. The ship crews aren't hunting for food, nor are they searching for sex. Everyone realizes that if the two ships were to engage, scores of people would be killed and crippled. The people on both ships would meet any standard test of sanity, and, to avoid a tragic event, all they need to do is nudge the rudder on either ship.

With all that in mind--and obeying an abiding instinct for self-preservation and love for their families – do the ship crews choose to avoid combat? No, they race toward each other, cannon at the ready. Cannonballs crunch into wood; grapeshot tears into flesh. As both sides expected, scores of people are killed and wounded. Arms and legs are blown off. Faces and chests are turned into pulverized meat. Bloodied intestines seep from mangled bodies. Finally, a salvo rips a hole in the hull of one of the ships, and it begins to sink. Shouting crew members below decks struggle to get through passageways choked with splintered wood. All of this was foreseeable and largely foreseen by everyone on both ships. And yet – and yet! – the two crews raced toward each other, eager for the fight.

**Third battle**

The lieutenant is not pleased with his situation. He and his platoon of 40 men had been ordered to occupy and hold part of a hill near enemy territory. This tactic is ordinarily prudent. It gives the defenders good visibility, with the advantage of shooting down, instead of up. But this hill has a problem: The ground is rocky and hard. The men curse as they slam the rocks with the edge of their entrenching tool. It was early evening when they occupied their section; they expect an attack that night.

The lieutenant has to move fast because there's a lot to do: Make sure each man is dug in and well-placed to defend his part of the hill...give coordinates to the battalion headquarters for potential artillery fire...position his machinegun crews to cover places where the enemy is most likely to advance...have Claymore mines set up in front of the hill...make sure each man is equipped with an ample supply of grenades...order a sergeant to watch the land in front with night-vision glasses...assign another sergeant to an M72 anti-tank weapon, not because he expects tanks, but rather for the intimidating bang of its burst ...walk the line to establish a command presence among the men, who are tense and silent as they wait for the onslaught.

Soon after midnight the first mortar shell explodes in the middle of their position. Someone yells "Medic!" The lieutenant calls for an artillery barrage, but the impacts fall behind the enemy in the foreground to avoid casualties from friendly fire.

The enemy infantry, which had crawled close to the lieutenant's platoon in the darkness, leaps up after the first mortar burst and charges into the defenders' position. The fighting is hand-to-hand as mortar and rifle fire incite more cries of "Medic!" These pleas go unheard – the defenders' heavy weapons platoon has started its own mortar barrage, aiming at enemy reserves behind the assault group. The din of battle is deafening.

The assault on the senses includes the smell of seared flesh as hot pieces of mortar shells tear into torsos, arms, and legs. Both attackers and defenders suffer heavy casualties. As the battle progresses, the defenders make the most of their preparations and manage to beat back the enemy and hold their position. In the eerie post-battle silence, the men exchange a few grim smiles. Two days later, the platoon and their whole company are ordered to abandon the hill they held. No reason is given.

**The sex of war**

Although these fictional battles differ in time, technology, and combatants, they have one major point in common: From the Stone Age to current news, nearly all the participants are male *Homo sapiens*, otherwise known as men. There are now, of course, a significant number of women in the military forces of many countries. But they rarely serve in combat infantry units. The basic fact remains: Whether in ancient Egypt or current Ukraine, the sex of people who kill others in massive numbers has always been male. Why is that? Why don't we see armies of women slaughtering each other? The answer will become clear as we move along.

**Purpose**

This book calls attention to one of the most important, yet under-recognized, problems in today's world: the obsessive inclination of men to wage war. Women, in general, do not share this inclination. It follows that if we want to reduce the number of wars, we need governments where women have at least as much political power as men.

War has always been a massively destructive force. But now it has become more dangerous to civilization than ever before. We still have the familiar world-ending weapons – biological, chemical, and nuclear. But we also have new kinds of weapons and new kinds of war, all with a huge leap in deliverability. Today's weapons can hit targets by means that did not even exist just a few years ago. One example: guided missiles that bolt through the atmosphere at many times the speed of sound, making them hard to track and intercept. In development: missiles and high-powered lasers that loiter in orbit, waiting for the command to strike. Already deployed: smart drones that communicate with each other and attack in swarms. These and other technologies – including self-actuating artificial intelligence – have made it quicker and easier to launch a war.

In charge of all this destructive power are *Homo sapiens* males, a species and sex infamous for its willingness to go to war, even when war is not needed to protect and strengthen national security. This book will explore and discuss the biological and evolutionary basis for that inclination – in our history, our hormones, and in the structure of our brain. It's important to note that this tendency is not a biological mandate; it's a propensity, something men are often inclined to do – but are not biologically compelled to do. Unfortunately, the factors that drive men to violence and war are often more powerful than those that lead to peace.

Compounding the danger posed by men's propensity for war is the phenomenon of widespread public trust, a naive belief that the alleged wise men leading advanced nations will "do the right thing," which ignores events like the invasions of Georgia, Somalia, Ukraine, and the 20-year battles in Afghanistan, Iraq, and Vietnam.

**Sources**
Using the methods of investigative reporting, the author draws on the work of dozens of scientists, scholars, and journalists in a broad variety of disciplines.

# Contents

# 1

## What Makes a Man?

A man used to be any adult with a penis and testicles. But it's not that simple anymore. The idea of what it means to be a man is changing. Or at least, that's what herds of pundits keep telling us. Shocking #MeToo stories compete for headline space with reports of "toxic masculinity." The American Psychological Association (APA) has issued Guidelines to help clinicians improve the health of boys and men. The document's Introduction spelled out a long list of problems that boys and men have to a far greater extent than do girls and women, including homicides and suicides. Citing a Department of Justice source, one of the APA's Guidelines points out that boys and men commit about 90 percent of violent crimes in the United States.

In *The Washington Post*, Christine Emba said that young men "are being outpaced by women surging ahead in school and the workplace...Cut loose from a stable identity as patriarchs deserving of respect, they feel demoralized and adrift, as if they don't know how to *be*" (italics by Emba).

Taking an opposing point of view, an article in the January 2019 *National Review* asserted, "With young men in crisis, The American Psychological Association wrongly declares war on 'traditional masculinity." A *New York Times* headline neatly encapsulates the hubbub: "Not Your Father's Masculinity."

So maybe it's time to take a stab at stating a short list of traits that characterize most of today's men. Doubtless subjective and idiosyncratic, this list at least provides some idea of what it means to be a man. As a species and sex, we seem to have these traits and desires:

**Have sex with a willing beautiful woman.**
If a willing beautiful woman is not available, then with a willing woman. This comes first in a list of male traits, because reproduction is vital to the survival of every species: The purpose of life appears to be more life. Reproduction is so important that in humans it's motivated in at least four different ways:

The first starts with the fact that sperm and some seminal fluid are produced in the testes and stored mostly in the epididymis, an elongated semicircular structure attached to the testes. As the quantity of stored semen builds, it creates a pressure that can be so intense as to be painful. This pressure may be necessary to propel the semen through the vas deferens, a long, twisted tube that winds up at the ejaculatory ducts, after visits to the prostate gland and the seminal vesicles, which produce more seminal fluid and add to the pressure.

Women may wonder why a man wants to have sex right now, this minute, with no more waiting while she does God-knows-what in the bathroom. This is the reason. The pressure from stored semen is similar to the sensation of an over-full bladder – something has to come out now, not later. When it finally does, from either organ, the man feels a great sense of relief. The sexual process is pressing, partly because it happens frequently – in teenagers and young men at least once or twice a day, every day, and two or three times a week in older men.

The pressure of accumulated semen accounts for most of a man's sexual desire. With lots of stored semen, thoughts of sexual activity spring to his mind unbidden, no matter how hard he might try to squelch them. Playing sports, especially rough sports, can help to manage all this desire, but biology eventually asserts itself. On the other hand, after one or more ejaculations, a man can focus on other things without a parade of distracting messages from his body. A woman who wants to keep a man would do well to keep him on empty.

All of which suggests that the pressure of stored semen, by itself, might be enough to assure frequent sexual activity. But no. Reproduction is so vital that evolution was not about to settle for a single incentive.

A second motivation for sexual activity is that it feels good. In fact, so profound and pervasive is the joy of sex with a willing woman that it may be the most intense brief pleasure a man can have. It doesn't take the place of long-term pleasures, like sharing a life with your spouse and seeing your kids do well. But the good feeling that comes with sex is no small thing.

A third motivation for men is the hope that his partner in sex will show obvious signs that she's feeling intense pleasure. This makes a man feel powerful, pleased with himself, and pleased with his mate. In most men it's a potent reason to have sex all by itself, which may explain why some clever women can get information out of men who would rather die than knowingly reveal state secrets.

A man (or woman) can have a happy and productive life without becoming a parent. But for most of us, having kids is the fourth motivation for sex and a large part what life is all about. It's also a huge amount of trouble and expense, and along the way, is likely to include sorrowful moments, as well as profound pleasures, including the times when a small person looks up at you and says, "daddy." The pain and the pleasure are what we call life, and in the end you think, "Yeah, it was hard. But worth it."

The reader may feel these observations make too much of sex as part of a man's life. Maybe. But think of how many prominent men have fallen because of sex. Leaders of countries and giant corporations, kings, senators, scientists, bishops, successful men in many fields – names we all know – tumble like bowling pins, one after another. It's been happening for thousands of years, and it will keep happening. For what? Does the quality of an orgasm depend on which vagina you have it in? All these men could have all the orgasms they want legally and within the bounds of marriage. So why do they risk everything they've worked for to have sex on the side? It may be because prominent men who carry on this way have a strong interest in power, and there are some women who are willing and able to make a man feel powerful – before sex, during sex, and after sex. Power, not orgasm, may be their main allure.

But for most of us, having sex presents an odd sort of contradiction. It's obviously very important, and yet, in a good relationship, it's a routine event, sort of like eating, eliminating waste, and sleeping: You don't think much about it. Until it doesn't happen.

This point raises serious questions about any religion that forbids its clerics to marry and have sex. The prohibition does not appear in either the Old Testament or the New Testament. It was invented by church officials and could be removed by church officials. Instead, at least one religion has been willing to live with its clerics abusing (and traumatizing) young boys and girls. The church denounces these crimes and sometimes punishes those who commit them. But it fails to do the one thing that would put an end to them: Allow its clerics to marry and live a natural life. Considering the sexual nature of men, a case can be made that the church should require its clerics to marry.

A curious sidelight to men and sex is the sexual behavior of older men. Does it exist? Yes, according to a 2010 *Science News* article:

> *At age 55, men can expect another 15 years of sexual activity, but women that age should expect less than 11 years, according to a study by University of Chicago researchers published online March 10 by the British Medical Journal. Men in good or excellent health at 55 can add 5 to 7 years to that number. Equally healthy women gain slightly less, 3 to 6 years. One consolation for women is that many of them seem not to miss it. Men tend to marry younger women, die sooner and care more about sex, the study confirmed.*

Why the difference in sexual interest and activity? The loss of interest by most older women is understandable. If a female is not making any eggs, she is not useful from a strictly reproductive point of view; the mechanisms of species survival would have no reason to preserve her sexuality. Why doesn't the same logic apply with equal force to men? Could it be the right stuff syndrome? The reader may remember Tom Wolfe's book, *The Right Stuff*. It reviewed the history of the early astronauts and their experiences as fighter pilots. Well into the book, Wolfe implicitly reveals what the right stuff is: The pilots who survived had the right stuff; those who died in crashes, didn't. That's egregiously unfair to many skilled pilots who crashed because of equipment failures or in combat. But Wolfe may have been expressing a common attitude.

Now, go back a few hundred thousand years and look at our situation. It probably wasn't pregnant or nursing women who speared large plains animals and brought back that critical nutrient we call meat. It was men, or mostly men. We know from studies of ancient broken bones that many were injured or killed as they tried to bring down fast-moving dangerous prey. Those who were not eliminated, even if they were too old to spear prey, were still valuable, because they knew how to hunt and kill. It's likely they were good teachers and leaders of hunting groups. Their sperm had survival value. They had the right stuff.

## A place for love?

In all this talk about sex, is there a place for love? Love that the poets sing about? Love that binds two people even more closely than sex? Yes. For one thing, sex with a woman you love can be the most satisfying form of sex. But we need to be clear about the difference between sex and love. For most men, simple sexual desire is wanting to ravish and ejaculate in a willing woman. That's all.

Love is what you *invest* in a person, pet, thing, or process. No investment, no love. There can be lust at first sight, but not love. Love takes time and effort. It includes the many things we do to show our woman that we care for her: little things like random kisses and hugs and bigger things like supporting her career and taking care of her when she's sick.

If a man invests more in his work than in his spouse, he probably loves his work more. When a woman says, "He loves his car more than he loves me," she may not be exaggerating. Can a man do both – love his partner and his work? Yes, but it takes an ongoing awareness of where and how he spends his time. It does no good to think, "Hey, I've put in the time and effort." No. The investment has to be continual.

Moving on to other traits that characterize most men:

## Control himself

When a man goes into combat, he is probably afraid. He should be; he may be killed or wounded in the next few seconds. But he controls his fear, partly because he inherits that ability from ancient ancestors who attacked and killed dangerous animals with crude spears. An ancestral hunter had to control himself well enough to do what was expected of him; he also had to avoid impulsive actions that involve needless risk. If you were in a hunting party, and happened to be in the best place to sink a spear into a lightly wounded wildebeest, your impulse may have been to run away. But you had to overcome that fear – or be relegated to the non-hunters: old men, most women, and children. This trait helps to explain why some men (and some women) climb treacherous mountains, ride giant waves, or put their last nickel into a startup – all while controlling the concomitant fear.

## Assert himself

A man needs to have a sense of what he's entitled to – at work, in marriage, as a parent, as a citizen. He has to be realistic about that sense, or he will look like a fool. But he cannot be shy about asserting it. He has a duty to do the best he can for himself and his family. In our distant past, a man had to assert himself to join a hunting party, then assert himself to get his fair share of the kill. Those who did that survived, reproduced, and passed their DNA on to the next generation. Biologically, nothing has happened to make this kind of behavior less useful, so it is still part of us, usually in the form of socially benign self-seeking. It is not directed at anyone; it does not involve malice or a desire to dominate others. It is simply acting to hold one's own among others and get a fair share of what society has to offer. In most men, it lives alongside efforts to help others.

## Gain and wield power

This is sometimes the dark side of male assertion. Many men – not just politicians, CEOs, and even academic department heads – will go to extreme lengths to get power and keep it. This trait may have its roots in a drive to be the alpha male, the male who leads or dominates other males and commands the most mating opportunities. The specific actions that produce a feeling of power vary greatly from man to man and for the same man in different situations. In advanced countries, the drive for power often shows itself as a quest for money and fame. It means, for example, working like a demon to climb a corporate ladder, make a scientific discovery, or win a literary prize. But it can also take the form of deliberately seeking dominance over others, often by bullying and random acts of cruelty. In its most extreme form, it leads to invasions and war. (Vladimir Putin comes to mind). This perversion of a basically constructive trait is so important, we will return to it.

## Care for his family

Loving and protecting his wife and children makes a man feel good. He enjoys making his partner happy, watching his children play a game, reading to them at bedtime, carrying a child on his shoulders – these are some of his greatest pleasures. He also takes pride in providing for his family financially, however hard it may be. There is a danger here, as well. We've all heard about men so intent on making money that they neglect their family. Finding a constructive balance can be challenging, but a man will come closer to that goal if he realizes it's part of his job as a husband and father and even more important than the commercial work he does.

## Win the respect of male peers

Men want to be seen as qualified to join the company of other men as a member able to accomplish the task at hand. This applies to how men talk, act, and even to how they dress and groom—not so much for women, but more for other men, to signal their status and the kind of team they're equipped to join. This trait may be most operative among men in combat. When combat veterans are interviewed, they rarely say they risked their life to kill or defeat an enemy; instead, they talk about their comradeship with other men in their unit and how they support and protect each other, even to the point of risking their life to save team members. That feeling is so strong that for some men, it's the main reason they re-enlist and willingly go back into combat. It's hard – maybe impossible – to find that kind of rapport in civilian life. One of the most important goals of a corporate or political leader is to instill a sense of comradeship among the people he or she supervises. The success of an organization may depend on it.

## Compete with other men

Men like action, especially competitive action: Playing a pick-up game of basketball or two-hand touch. Sprinting to the next paintball structure without getting hit. Shooting a three-pointer while closely guarded. Even when not competing against others, men like to compete against themselves: Can I climb that mountain? Can I do this stunt on my skateboard? Can I run that rapids? For many men the competition may be sedentary, but just as intense: Can I trap his queen? Can I beat the market? Can I win this video game?

Our history as a species suggests it's in our DNA to prove ourselves and, especially, to prove ourselves better than other men. The motivation may come from our distant past when the man who could spear more animals and bring back more meat than other men probably had the best mating opportunities and the most children. Many women also like competitive action, but there's a difference: Women, until they're about 50, prove themselves as females once a month, every month, and may give birth to children, the ultimate proof of sexual identity. Men don't have those advantages, so they do other things, perhaps more intensely than women.

## Take risks

Men – especially young men – tend to like physical risks, like hang-gliding, American football, or white-water kayaking. But the risks can also be psychological, like trading commodities, playing high-stakes poker, or starting a business. It's often said that people take risks for the adrenaline high. There's no question that the feel-good rush is part of the motivation. But there may be a deeper drive. When our ancestors hunted large animals for food, it took a lot of courage to get close enough to a dangerous animal to kill it or force the animal into a path where others could kill it. As with some other traits in the list, the men who took those kinds of risks (and lived) probably had more opportunities to pass on their DNA with willing women. Their genes live on in men willing to take reasonable risks.

A related trait is the ability to endure pain. Part of many hunts probably involved the need to keep going through painful situations, physical and emotional. That point explains why many tribal societies have painful initiation rites for boys when they reach the age of 12 or 13. The boys must prove to themselves and to the tribe that they can cope with any situation that may come up during a hunt. Some of the major religions have ceremonies that recognize the coming-of-age idea, without painful or dangerous physical acts.

## Build things

Men like to build things, make things, invent things. Little or big. And for some men, the bigger, the better. Many women have a similar urge. But in women, it may be mainly a desire to produce things that make life better – an extension of their drive to nurture. In men, the process itself may be the driving force. Men can't conceive and give birth to children. Building things is as close as they can get to showing what they can produce. In ancient times, men had to build or improve shelters to attract and keep women. To feed themselves and their brood, they had to invent and make tools and weapons. Nothing in evolution has happened to lessen the value of those kinds of inclinations and skills. Today, they may take a different form, as in constructing a computer program or inventing a new kind of financial instrument. But the primal urge is the same.

## Engage in violence and war

As already stated, the history of wars, weapons, and interpersonal violence indicates that men in general have a propensity – not a compulsion – for violence and war. We can avoid war, but in many situations, we are all too willing to shoulder a weapon, wield a drone-controlling computer, or order others into war. We will explore this trait in the pages ahead.

## Other men

Biological research shows human sexuality is not binary; it exists on a spectrum from female to male. As we will see later, scientific studies tell us much of human male sexuality is determined by events that occur in the uterus or within a few months after birth and again at puberty. This fact means a person born with a male body may grow up to feel like a female; a person born with a female body may eventually feel like a male. Deliberate choice has little or nothing to do with it. Males who prefer sex with other males are not "queer" or depraved; they are simply less frequent in the gene pool, somewhat like people with blue eyes. And since they had not the slightest choice in their preference, they cannot logically be seen as immoral or sinful.

# 2

## The Wrong Question

Philosophers and historians have been asking what causes war for centuries, and yet there does not seem to be a universally accepted answer. The reason is that people have been asking the wrong question. It's not a case of *what* causes war – it's "*Who* causes war?" When we speak of war, we typically mean one country fighting another, with no overt distinction regarding the sex of each country's participants. This is misleading. As already stated, it is almost always men who start wars and men who fight them. Not women.

In the *Encyclopedia of War*, American University professor Joshua Goldstein states, "Historically, of the untold millions of combatants in the world's wars, more than 99 percent have been males." In a 2008 *New Scientist* article, political scientist Rose McDermott tells us, "There is something ineluctably male about coalitional aggression – men bonding with men to engage in aggression against other men." In a 1998 *Foreign Affairs* article, political scientist Francis Fukuyama refers to work by Lionel Tiger:

> *Nearly 30 years ago, the anthropologist Lionel Tiger suggested that men had special psychological resources for bonding with one another, derived from their need to hunt cooperatively ... Tiger was roundly denounced by feminists at the time for suggesting that there were biologically based psychological differences between the sexes, but more*

*recent research ... has confirmed that male bonding is in*
*fact genetic and predates the human species.*

Which raises a question: How to explain this persistent dedication to a brutal and horrific way of settling differences, by a species with enough intelligence to unravel the mysteries of the universe?

The answer starts with the need for males to compete for mating rights, which often requires assertion of dominance, some form of aggression, or in many species, physical combat. This is one of the ways evolution works to improve the survivability of a species. The winners of these contests tend to have genes better suited for the local environment than do the losers. In humans, as well as in chimpanzees, this trait may promote coalitionary attacks on neighboring groups. More on this point later.

## The way it was

Many scholars have pointed out that our ancestors were ill-equipped to become a species that would dominate the earth. Compared with other predators, like the big cats, bears, bone-crunching hyenas, and even most of our prey species, we were smaller, weaker, and slower. Worse, we lacked the wicked canine teeth, pounding hooves, and formidable horns of our four-legged prey. We had a big brain, but the Neanderthal brain was even larger. How, then, were we able to rule the earth?

One answer comes from the eminent Harvard professor, E. O. Wilson, who invented a science called sociobiology, which holds that social behavior developed mainly through biological evolution, not culture. Wilson states the principles of this discipline in an illustrated book entitled *Sociobiology,* a monumental work that defined a new way of analyzing human behavior – how our evolutionary past may have helped to cause traits we see in humans today.

However valid that thesis may be, at least one of the excellent drawings in *Sociobiology* is open to question: Covering two pages, it shows eight early humans trying to defend the carcass of a huge, prone animal from three saber-tooth cats and three hyenas. It can be argued that the drawing, though exquisite in detail, portrays at least two dubious situations:

The first is that it shows the eight men acting separately from each other. It's hard to believe this would have been the case. More likely, the men would have formed a line with their back to the carcass and stood roughly shoulder-to-shoulder, clearly indicating they would act as a unit. This is highly probable, because that's what men have always done in battle.

The second point is a fault, if the intention was to show a typical situation. The picture depicts a gross imbalance in lethal power: Two men are throwing a rock, and one wields a short, fat stick. The rest are waving their arms overhead. Consider the natural ability of the attackers to inflict harm versus that of the defenders: Each of the cats can jump at least 20 feet, then tear its prey apart with enormous canines. The hyenas usually attack in a coordinated way, from the front, back, and sides, then bite with the most powerful jaws of any mammal. It appears that all the humans who chose to stay would be dead or dying within seconds. They would probably have met this fate, even if they had worked together. So how did we, often enough, claim the carcass?

## Working together, with weapons

Picture the same scene with two crucial differences: The men stand shoulder-to-shoulder, acting as a unit, and this time each holds a bare tree branch about seven feet long, sharpened to a penetrating point – they have spears. The men point their weapons at the attackers and make jabbing movements to emphasize the inherent danger of the points. They make guttural noises in unison to add to their sense of menace. The cats and hyenas have never seen anything like this before and know they can't afford to get hurt. The odds have shifted. The men still face a high-risk situation, but now they have a chance to prevail.

## A turning point

The spear, in fact, was the first major step in human domination of the earth. According to *Scientific American* author Kate Wong, the earliest effective spears may have been made by proto humans called *Homo heidelbergensis,* generally thought to precede both Neanderthals and humans. *Heidelbergensis* is thought to be the maker of the Schoeningen spears. These are 10 wooden spears that were found in 1995 in Schoeningen, Germany. Although more than 300,000 years old, they are not crude weapons. Measuring about seven feet in length, the spears were made from a strong variety of spruce tree, with the forward section fashioned from the base of a tree trunk, where the wood is the hardest. Each spear balances in the hand one-third of the way from the front, as do modern javelins. Fashioning such weapons obviously required a high level of craftsmanship and cognitive abilities. There is little doubt about the effectiveness of these spears and the males who threw them. A 2015 *Archaeology* News Brief tells us, "Thousands of pieces of horse, elephant, and deer bone were also found at Schoeningen."

## The arms race heats up

Wooden points can be lethal, but they tend to have limited penetration power, so our ancestors must have searched for a better weapon. What they developed represents a major advance in the history of weapons: Extensive research by University of Cape Town professor

Jayne Wilkins showed that proto humans learned how to attach a piece of sharpened stone or bone to the end of a tree branch – and did it about 500,000 years ago, far earlier than had been thought likely. The various essential steps evolved over many thousands of years. First, the stone itself may have required heat-treating to make it harder and easier to chip. The final shape of the spear point had to have a sharp but sturdy front end, and the other end had to made flat and thin enough to fit into the split end of a suitable tree branch. Then, for additional strength, the spearhead had to be wrapped tightly with animal gut and cemented with some sort of gum or resin. The total process required foresight and deft handwork based on remembering which chipping and binding techniques work and which don't. According to some scientists, the process itself may have stimulated brain growth in meat-eating *Homo* species.

This invention gave the spear greater penetrating power; it could now inflict serious wounds, even when thrown from a distance. That's why it prevailed for many thousands of years. The atlatl, a leveraged way of throwing short spears, did not appear until about 18,000 years ago. Hard-metal points on spears would have been a big help, but they were not available until the Bronze Age, which started only 3,300 years ago. The bow-and-arrow may have been invented as early as 70,000 years ago but was not widely used until about 3,000 BCE.

> All of which means that for nearly 300,000 years, primitive spears were all we had to satisfy our need for meat to eat and furs to wear.

With these facts in mind, we can infer that ancestral hunting was hard and dangerous work. To deal with those threats, our ancestors probably needed five sets of abilities:

- They had to have enough intellect to design and craft effective spears, without the aid of modern tools.
- They had to be able to run far, sweat profusely, and throw spears with power and accuracy.
- The hunters required a plentiful supply of the so-called male hormones – testosterone and other biochemicals that promote aggressive action in the face of imminent harm.
- It's likely the hunters needed to work closely as a team to track, wound, and kill their prey.
- The hunters probably needed spoken communication and dominant males to coordinate their efforts.

Young adult females, typically either pregnant or nursing, probably learned to prefer males who brought back meat to the group. It's likely they had to be assertive to make sure they got an adequate share for themselves and their children, but there was no point to physical assertion; males were typically bigger, stronger, and far more prone to violence. So females had to find other ways to get what they wanted. Fortunately, they were only partially dependent on male largesse. Many hunts for large prey were probably unsuccessful, so it's likely that females gathered plant food and hunted small animals to augment their group's total diet. Aided by their own set of hormones, mothers had to be persistent in caring for children; it took more than a decade for human children to become independent.

With an uncommonly large and complex brain, *Homo sapiens* lived in scattered bands on the African savannahs and the South African coast, until about 75,000 years ago, when the Toba super volcano in Sumatra sent up a dark cloud of ash that may have damaged life on much of the earth. Scientists are still debating the issue, but this event could have been one reason some humans around that time left Africa for the Mideast and parts of Europe. (There is some controversial evidence suggesting that the human population shrank to a very small size about 93,000 years ago.) Eventually, of course, the enduring combination of aggressive hunters and persistently nurturing mothers was so successful that humans were able to populate most of the earth.

## The fate of other *Homo* species

Scientists debate the number of other *Homo* species, with estimates ranging from 9 to 21, a few of which lived at the same time as *Homo sapiens*. What happened to them? Why aren't we sharing control of the earth with other species? Scientists have several competing theories. One is that we disposed of *Homo neanderthalensis, Homo erectus, Homo heidelbergensis,* and other species by outcompeting them for resources.

Another prominent belief is that there were several causes, including the possibility that we killed them. It may not have been a giant leap for human males to evolve from throwing spears at quadrupeds seen as food to throwing spears at bipeds seen as competitors: "Our water hole, not yours. Our hunting grounds, not yours. Our females, not yours."

Technical note: The earliest appearance of *Homo sapiens* used to be reckoned at about 200,000 years ago. But in 2016 and 2017, researchers from the Max Plank Institute found fossils in Morocco that were dated to 315,000 years. Not every paleontologist agrees, but many believe these ancient bones are those of archaic *Homo sapiens*.

## Different work, same men

The next major event in the human story began about 10,000 years ago, when humans are known to have started farming and animal husbandry. It's true that tilling the soil, planting seeds, and raising animals require a different set of attributes from those needed for hunting. But the requirements of farming would not have changed the DNA of human males, for at least two reasons:

- With rare exceptions, actions taken by an animal during its lifetime cannot change its DNA.
- The entire time since humans started farming until today amounts to less than four percent of the known time we've existed as a species – a mere blip as evolutionary time is reckoned.

## Why are we violent?

The late arrival of hard metal and farming means that for nearly 300,000 years, natural selection and sexual selection continually reinforced the physical, hormonal, and organizational qualities needed by poorly armed hunters of large animals. Judging by how evolution usually works, successful hunters probably mated more often and had more children than did non-hunters or second-rate hunters. Even at a conservative 30 years per generation, that's 10,000 generations during which the DNA of successful hunters was passed on, again and again.

That process suggests the genes of successful hunters have been rooted in the genome of male *Homo sapiens*. This may be the central fact that explains the combative nature of men, who, even today, seem obsessed with cooperative group violence and the incessant development of ever-more-deadly weapons.

The available evidence strongly suggests that violence and war are defining characteristics of human males, operating as a massively lethal atavism. This dismal fact presents a disturbing paradox: Men no longer need to kill dangerous wild animals with primitive weapons, so why is the hunter-killer DNA still present in today's males? The likely answer is that nothing has happened to weaken the efficacy of aggressive, risk-taking men. The qualities of pre-Stone Age hunters are still useful for achieving success in competitive economies. Women are still attracted to men who seem to have these qualities, and society still rewards such men. Consider the kind of men we admire and support:

- *Football players*: strong, aggressive, willing to risk life and limb. The top-earning NFL player earns $50 million a year.
- *Action heroes*: strong, aggressive, and seeming to risk life and limb (in movies). In 2022 the highest paid action hero actor earned close to $100 million.
- *Successful entrepreneurs*: financially strong, aggressive, willing to risk huge sums of money. Nine of the richest U.S. men have more than $100 billion in assets – some, a lot more.

Remembering that a society gets what it pays for, compare those sums with the income of most scientists, male or female: According to salary.com, their average annual salary in the U.S. runs to about $175,000 – a tiny fraction of the top-earning actor's income in one year. As a society, we place a higher value on entertainment action heroes than on people who help to improve the quality of life.

There are other explanations for the violent behavior of *Homo sapiens* males. Some say it's a battle for resources; others say it all comes from human culture. But neither explanation accounts for what soldiers go through on a battlefield. When bullets snap a few feet overhead, mines lurk underfoot, and mortar shells burst on all sides, every human instinct is to run away. But infantry holds its ground or moves ahead, into the fire, as Ukrainian soldiers are doing today. You don't get that kind of behavior just from reaching for resources or conforming to culture. It has to come from something more fundamental. In the pages ahead, we will see what that is.

## Equal power for women

All of which leads to the central point of this book: It has become imperative that women assume at least as much political power as men. Women know how to be assertive, even aggressive, without getting physical; they are far more inclined than men to use non-violent means to attain their goals. That particular trait is urgently needed in today's world. The reason is not just the hope of reducing the number of traditional wars; it's also because the threat of cataclysmic wars has increased to a level that may be the highest in history. Threatening the world today are not only the old Cold War weapons but also new weapons and new kinds of war, with their danger amplified by the power of artificial intelligence.

## The problem

It can be argued that the message of this book is needlessly extreme, as expressed in a long-standing assumption: "Yes, it would probably be good if women had more political power, and it could be useful for federal officials to realize that some men have a bias toward using war as an instrument of policy, but there's no need for a disruptive change. When sufficiently informed, men, or the current mix of men and women, can handle whatever comes up."

However much we'd like to believe that assumption, it's probably wrong. If rule by men hasn't stopped or even slowed the pace of wars for thousands of years, why should we believe it will work in the future?

The problem is that the idea of having women assume equal power threatens many men, including government officials. As we'll see in the pages ahead, men are all about power. For many people – some females, as well as males – a man's power defines the man. Ceding a share of that power to women – enough to give them an equal voice in national affairs – will probably be resisted as much in the future as in the past.

## Confronting the resistance

The need to challenge that opposition explains why much of this book focuses on the combative nature of men and why, acting alone, they cannot be trusted to avoid needless combat and war. Yes, there are exceptional men who can transcend their inclination to use war to pursue national goals. But are these the men most likely to seek and hold high office? It's a question to which the world needs a better answer than "Let's hope for the best."

# 3

<hr>

## Perpetual War

*"The story of the human race is war. Except for brief and precarious interludes, there has never been peace in the world, and before history began, murderous strife was universal and unending."*
Winston Churchill, The Week, 3/17/23, from Commentary

We like to think of being at war as an exceptional time. It's not. War has been the rule; peace, the exception.

Pulitzer Prize journalist, Chris Hedges, makes this point in his book, *What Every Person Should Know About War:* "Of the past 3,400 years, humans have been entirely at peace for 268 of them, or just 8 percent of recorded history." If accurate, those figures indicate that humans have been at war 92 percent of that time. Indeed, scientists have found evidence of human mayhem that occurred in prehistoric times:

In his book *Where Does Violence Come From?* scientist-author Bernhard Bogerts comments: "The fact that warfare was part of the repertoire of prehistoric hunter-gatherers is proven by numerous excavations...At a site in Sudan, remains of 59 people from the period 14,000 to 12,000 BCE were found with flint points in their skeletons."

In a 2016 *Smithsonian* article, author Brian Handwerk, cites the ancient slaughter of 27 people:

*Even nomadic hunter-gatherers engaged in deliberate mass killings 10,000 years ago ... The battered skeletons at Nataruk, west of Kenya's Lake Turkana, serve as sobering evidence that such brutal behavior occurred among nomadic peoples, long before more settled human societies arose.*

The article includes a quote by Marta M. Lahr of the University of Cambridge:

*The injuries suffered by the people of Nataruk – men and women, pregnant or not, young and old – shock for their mercilessness ... what we see at the prehistoric site of Nataruk is no different from the fights, wars and conquests that shaped so much of our history, and indeed sadly continue to shape our lives.*

Note: This chapter is thick with cold, hard numbers. We need them to document the extent of the human misery they imply: a river of tears for those who can no longer cry.

## 28 U.S. wars

The military history of the United States provides more recent examples of perpetual war. According to Martin Kelly on thoughtco.com, Americans have been involved in 28 wars since 1775. The conflicts included the wars in Vietnam and Afghanistan, both lasting 20 years, the eight-month war in the Persian Gulf, the one-month war in Panama, and many other military actions, including the unwarranted invasion of Iraq, leaving almost no time during which the United States was entirely at peace.

Meanwhile, other nations have also been waging war. An *Encyclopedia Britannica* article covers the "8 Deadliest Wars of the 21st Century." The Second Congo War is cited as "far and away the deadliest war of the 21$^{st}$ century." The remaining seven include combat in Syria, Sudan, Iraq, Afghanistan, Nigeria, Yemen, and Ukraine. Many millions of people died in these wars (three million from all causes in the Congo alone). Millions more were rendered homeless, driven to become refugees, or died of disease fostered by the wars. This does not count the millions more who died in wars that were not among these eight.

A country's citizens may underestimate the frequency of wars, because governments often soft-pedal their military actions. Writing in *History News Network*, Paul Lovinger makes the point that even major wars can occur without a formal declaration:

> *Congress tries to end the President's [Trump's] support for the Saudi-led slaughter in Yemen. In Syria too, U.S. forces engage in undeclared bloodshed. And, though Congress never declared war on Afghanistan, America's struggle there nears eighteen years [20 years by 2021]. Many young Americans have never known peace.*

## The cost in lives lost and damaged

Given the high frequency of wars, a distant observer might think that waging war must be relatively cheap. We know that's not true, but just how untrue may be an eye-opener.

The Watson Institute at Brown University has taken on the task of investigating and reporting the true cost of wars and military spending. It describes itself as an interdisciplinary research center with the mission of promoting a just and peaceful world through research, teaching, and public engagement. Here are some research findings on the human cost of war that were reported by Watson. Their June, 2023, report tells us:

*Military spending makes up a dominant share of discretion-ary spending in the United States; military personnel make up the majority of U.S government manpower; and military industry is a leading force in the U.S. economy.*

Taken together, these facts indicate that defense contractors have enormous economic power, in Congress and throughout the U.S. economy.

### The human cost in Iraq and Syria

According to a March, 2023, report, "Between 550,000 and 580,000 people have been killed in Iraq and Syria since the U.S. invaded Iraq in 2003 – and several times as many may have died due to indirect causes such as preventable diseases." Not to be overlooked, more than seven million people from Iraq and Syria are currently refugees, and all that does not begin to count the displaced people in Gaza, Myanmar, and other countries.

### The human cost in the Mideast

A unit of the Watson Institute reports that 241,000 people have died as a direct result of the war in Afghanistan and Pakistan. Referring to several Mideast countries, a 2020 U.S. Department of Defense Status Report reveals that nearly 7,000 Americans have been killed since 2001; an April, 2019, Congressional Research Service post reports nearly 53,000 wounded.

### The human cost globally

Elsewhere, the numbers are much larger. According to the Watson Institute, more than 801,000 people worldwide have died as a direct result of fighting just since 2001. Of those, more than 335,000 have been civilians. In addition, 211 million people have been forced to become refugees.

Compared with the finality of the number of people killed, the number wounded gets far less attention, but is still a grim statistic. Many people who read the word "wounded" may think of the typical movie scene where a damaged soldier is told by his buddies that he will be "all right" and that he's "going home." Going home to what? In all too many cases, it means a life without legs or hands, or with damaged internal organs, and pain, mental and physical, that may continue for the rest of his life. (Of course, female military personnel are also wounded, as Senator Tammy Duckworth reminds us, just by doing her job without her legs, lost when the helicopter she was piloting was shot down in combat.)

## The cost in U. S. dollars.

### Iraq and Syria

These figures come from the Cost of War project at Brown University: "The estimated U.S. costs of war in Iraq and Syria will reach $2.9 trillion dollars by 2050." (Even though the war is over, costs like payments to veterans, interest on the debt, and U.S. Department of Defense Operations continue for decades.) Looking back a little farther, the Cost of War Project reports:

> *Since late 2001, the United States has appropriated and is obligated to spend an estimated $6.4 trillion through Fiscal Year 2020 in budgetary costs related to and caused by the post-9/11 wars—an estimated $5.4 trillion in appropriations in current dollars and an additional minimum of $1 trillion for U.S. obligations to care for the veterans of these wars through the next several decades.*

It's tempting to imagine the many benefits that could be achieved if only some of that money – say, $1.5 trillion – were spent on things like repair of our deteriorated infra-structure, more pre-school facilities, control of global warming, and more accessible health care. It's not likely to happen: That kind of social benefit spending would not satisfy the male affinity for violence and war.

*Afghanistan and Pakistan*

In April, 2021, President Biden called for an end to the U.S. involvement in Afghanistan, an act that may have prompted The Costs of War project to publish an update on the cost of the war in Afghanistan and Pakistan:

> *Since invading Afghanistan in 2001, the United States has spent $2.3 trillion on the war, which includes operations in both Afghanistan and Pakistan.*

These figures do not include all indirect costs, such as the tens of thousands of lives lost because of the side effects of wars.

## U.S. the big spender

It is said that the U.S. must spend heavily on arms, because our enemies are spending even more, or could spend more. Not to dispute the need for some level of arms spending, the actual numbers suggest spending programs that lack adult supervision:

According to a 2023 *Axios* article, the U.S. in 2022 spent $877 billion on its military – more than the next 10 nations *combined* and more than 10 times the amount spent by Russia and three times the amount spent by China.

Even a single sector of our military spending can be wildly expensive. A 2020 *New York Times* article reports that the Pentagon's nuclear weapons modernization effort is on track to surpass its $1.2 trillion price tag over the next three decades.

## War by any other name

The frequency and cost of war is often underestimated by how war is defined. A few anthropologists are fond of insisting that wars are infrequent, because (they say) war means one nation fighting another, a hierarchy of commanders, and thousands, or at least, hundreds, of people on each side. This reduces the apparent number of wars by definition and supports their case. We see this argument most often in attempts to show that isolated populations do not fight wars. But scientists who have devoted most of their career to studying wars report that isolated tribes do fight wars, often with high casualty rates. The scale is different, but the essence is the same: two groups of combatants who have never met, trying to kill each other.

It's true that a fight between two persons (or two families) who know each other does not constitute war. But to the infantry soldier in a foxhole, or the pilot of a Mach-2 fighter plane, war can be as small as a one-on-one encounter. Even large-scale battles often play out in a series of small-scale confrontations. As a typical example, consider a squad of soldiers charged with attacking a four-man machinegun nest. The combatants have never met, but somebody lives, and somebody dies. That's war.

## Definite loss vs. hoped-for gain

The U.S. has spent hundreds of thousands of lives and trillions of dollars in a series of wars, most notably in North Korea, Vietnam, Afghanistan, and Iraq. Having spent a lot in lives and treasure, did we get a lot? Has even one of those countries become a functioning democracy? Did we gain valuable resources? Did we prevent additional military action? Did we deter hostile countries like Russia, China, and North Korea from taking aggressive actions?

This is not to say that all wars are pointless. We certainly had to fight World War Two, for example, and since we don't know what kind of war may come, don't we have to prepare for the worst? The answer depends on how we define "prepare." So far the U.S. has shown a desire to have the most advanced form of almost every class of weapon and to have it in over-whelming quantities. Is that the best policy? Do we have to flaunt an ability to totally destroy an enemy country? Or would it be more cost-effective to show that any attacker would be severely damaged? The answer could add a trillion dollars to the amount we can spend on education and the environment

# 4

## The Supermarket of Death

The current danger to civilization is not just that men are drawn to war; it's that we seem impelled to make war ever more lethal. Otherwise, we would still be fighting with clubs and spears, instead of weapons that can kill a hundred thousand people in a single strike.

On January 23, 2020, John Mecklin, editor of the *Bulletin of the Atomic Scientists*, announced that the Doomsday Clock had been set closer to presumed apocalypse than ever before in its 70-year history. The Doomsday Clock is a graphic device that stands in the lobby of the *Bulletin* offices in Chicago. It was created by scientists in 1947, two years after Hiroshima, as a news-worthy way to convey how close we are to man-made destruction of the world.

The hands on the Doomsday Clock are set periodically by the Bulletin's Science and Security Board, which consists of about 20 scientists and other experts who consult with colleagues and the Bulletin's Board of Sponsors, which includes at least a dozen Nobel laureates. In 2023 the Clock was set at 90 seconds to midnight, signaling a perilous situation.

Mecklin's 2020 announcement demands worldwide attention. It states, in part:

> *Humanity continues to face two simultaneous existential dangers – nuclear war and climate change – that are compounded by a threat multiplier, cyber-enabled information warfare, that undercuts society's ability to respond. The*

> *international security situation is dire, not just because these*
> *threats exist, but because world leaders have allowed the*
> *international political infrastructure for managing them to*
> *erode ... Civilization-ending nuclear war – whether started*
> *by design, blunder, or simple miscommunication – is a gen-*
> *uine possibility. Climate change that could devastate the*
> *planet is undeniably happening.*

There is a standard response to this kind of warning: "Yes, there is some level of threat, but it hasn't reached the point where radical action needs to be taken. After all, during the Cold War, The Soviet Union and the United States brandished weapons that could destroy an entire city with a single bomb, and nothing happened."

The argument sounds good – and we want to believe it – but it's dangerously misleading. There were, in fact, times when the world did come close to nuclear devastation. According to an *Atlantic* article, one incident occurred in 1983, when a satellite early warning system told a Soviet official to launch missiles in retaliation for a perceived strike on the Soviet Union by five U.S. ballistic missiles. The official was supposed to obey instantly. But he hesitated and finally decided it must be a mistake. The cause turned out to be not U.S. missiles, but a flaw in the Soviet warning system. The Soviet official who made the correct decision, Stanislav Petrov, may have averted a nuclear war.

During the Cuban Missile Crisis, war was much closer than the American public realized. According to history.com, the U. S. Strategic Air Command was ordered to Defense Condition 2 (DEFCON 2), the highest level it has ever reached. Over Europe, American bombers were in the air 24 hours a day, each ready to deliver a nuclear bomb. Soviet submarines armed with nuclear weapons moved into Caribbean waters as U.S. warships patrolled the area, ready to enforce a "quarantine" that had been imposed by President Kennedy. A U-2 reconnaissance plane was shot down over Cuba, and the pilot killed. War seemed imminent.

According to a 2012 *Christian Science Monitor* article, Robert Kennedy revealed in his memoires that the Joint Chiefs of Staff strongly and unanimously urged a military invasion of Cuba. Defense Secretary Robert McNamara argued for a blockade, instead. So strong was the urging for a strike that Robert Kennedy informed President Kennedy (and the Russians) that a U.S. military coup could occur if the U.S. did not invade. Fortunately for the world, President Kennedy decided to take McNamara's advice. Both the Soviet Premier, Nikita Khrushchev, and President Kennedy played for time and, after fraught negotiations, reached agreement: The U.S. promised not to invade Cuba and would remove its missiles from Turkey; the Soviet Union would withdraw its missiles and bombers from Cuba. The stand-off ended peacefully, but for 13 days the world was poised on the brink of nuclear cataclysm.

In a 2023 edition of *The Washington Post*, the editorial board pointed out, "The world is sliding into a new age of nuclear risk – in which miscalculation or accident could lead to catastrophe."

## Robotic warfare

After the Cold War, many people in government had only minimal fears that the major powers would attack each other with nuclear weapons. The consequent devastation on both sides was simply not acceptable; fear of global war subsided.

Things are different now, in several ways. In the February 2020 issue of *Scientific American*, an article by Noel Sharkey brings the world up to date (and perhaps up short, as well). Sharkey is a professor emeritus of artificial intelligence and robotics at the University of Sheffield in England. The main thrust of his article is that robotic warfare substantially aggravates the already-dangerous state of international military confrontations. The article begins by reporting that "a swarm of 18 bomb-laden drones and seven cruise missiles" attacked Saudi oil facilities in 2019 with devastating effect. Saudi Arabia's oil production was cut in half, and the global price of oil rose. Yemen's Houthi rebels claimed responsibility. The article does not explain how a mere rebel group could have launched such a sophisticated attack. But it brings up another disturbing trend in international power struggles: war by proxy. A big nation that doesn't want to get its hands dirty supplies a small nation or rebel group with advanced weapons and logistical support. This strategy adds to the danger of war, because it promotes a proliferation of relatively small groups with considerable destructive power.

Sharkey describes a range of air, ground, and naval robotic weapons, some of which can be programmed to make tactical target decisions on route to a strike. The most advanced are hypersonic missiles that can reach speeds of 6,000 to 14,000 miles (21,000 kilometers) per hour. *The New York Times* reported there is no known defense against advanced hypersonic missiles. By the time a target's detection system sees such missiles, it could be too late to intercept them. Compounding the threat, such weapons are maneuverable in flight, so they're hard to track and target.

Moving on to other weapons, Sharkey's article refers to "autonomous drones that cooperate like wolves in a pack, communicating with one another to choose and hunt individual targets." Sharkey's most troubling point may be how easily the systems that control robotic weapons can be subverted: "In reality, protecting against disruptions by the enemy will be extremely difficult, and the [results] of these assaults could be dire." Jamming, decoys, and spoofing can mislead sensors and make it impossible to control automatic weapons after they've been deployed.

Update: On December 1, 2020, the *Bulletin of the Atomic Scientists* published an article reporting that drones have entered a "second drone age," which is "marked by the uncontrolled proliferation of armed drones, the most advanced of which are stealthier, speedier, smaller, and more capable of targeted killings than a previous generation." The article also reports that more than 100 nations have military drones.

Supporting the *Bulletin* warning, a June 2021 article in *The Week* magazine emphasizes the lethal risks of drone proliferation:

> *The U.S. drone program steadily expanded in Afghanistan, Iraq, and Pakistan, and by 2014 the Air Force was training more drone pilots than airplane pilots ... Experts worry that the proliferation of UAV's [Unmanned Aerial Vehicles] could make bloody conflicts more common, because countries that are reluctant to start a war and risk their soldiers' lives may not hesitate to send in drones.*

The Sharkey and *The Week* articles focus on robotic weapons, ignoring another disturbing fact: At least eight countries are known to have nuclear weapons. It is widely believed that Israel also has such weapons, and Iran may be able to produce them within a relatively short time. Two of the eight countries, India and Pakistan, have fought two wars with each other, and tensions have continued.

Scientists tell us that if too many nuclear weapons explode within a short time, there is danger of a nuclear winter, during which smoke and ash from firestorms would block out the rays of the sun, make the earth much colder, and interfere with the photosynthesis that plants, and plant-eating animals, need to stay alive.

A date to remember: On January 22, 2021, a decades-long campaign by the International Campaign to Abolish Nuclear Weapons

(ICAN) came to fruition: Ratified by 50 nations in 2019, the UN treaty to ban nuclear weapons has entered into force. According to an ICAN press release, the new agreement "prohibits nations from developing, testing, producing, manufacturing, transferring, possessing, stockpiling, using or threatening to use nuclear weapons, or allowing nuclear weapons to be stationed on their territory."

ICAN Executive Director Beatrice Fihn, said this about the new agreement:" The treaty is a major step forward, but its effectiveness will depend on the extent to which men alone continue to dominate international relations and war policy, rather than coalitions in which women and men have equal political power."

The phrase, "coalitions in which women and men have equal political power" is a key idea in this book and one we will examine in more detail.


## Cyber insecurity

We've seen that enemies can take control of a target country's robotic weapons; perhaps equally disturbing is that they can also cripple a country's infrastructure, from inside its own borders. In the December, 2019, issue of *Scientific American*, science author Paul Tullis describes this threat:

> *The U.S. Air Force maintains 31 Navstar satellites that send radio signals to GPS receivers worldwide...Although we think of GPS as a handy tool for finding our way to restaurants and meetups, the satellite constellation's timing function is now a component of every one of the 16 infrastructure sectors deemed 'critical' by the Department of Homeland Security.*

The trouble is, the satellite signals are weak and easily jammed or "spoofed" by stronger signals. (Spoofing refers to broadcasting false information to GPS receivers.) Several experts told Tullis that such attacks could "severely degrade the functionality of the electric grid, cell phone networks, stock markets, hospitals, airports, and more – all at once, without detection." Seaports and ships are also heavy users of GPS systems. One expert said, "A coordinated spoofing-jamming attack against various systems in the U.S. would be easy, cheap, and disastrous.".

## A warning from Robert Gates

The threat to the U.S. GPS is underscored by Robert Gates, former director of the CIA, former Secretary of Defense, and one of America's leading authorities on national security. In his 2020 book, *Exercise of Power*, Gates warns:

> *While I believe the big powers—above all, the United States, China, and Russia—would refrain from any such large-scale attacks on each other short of a major war, the same cannot be said for North Korea or Iran if faced with a threat to the regime. Nor can any country expect restraint in the use of cyber threats by nonstate entities, such as terrorist groups, should they acquire that capability.*

However dire that may seem, we can still reasonably ask, "How likely is that to happen?" Gates provides a partial answer: "A conventional military attack is guaranteed to provoke prompt retaliation. Proving beyond doubt the origin of a cyberattack however, is both difficult and time-consuming." This fact heightens their appeal as a weapon. Weighing all these points, an enemy of the United States might well decide it can inflict the most damage with the least risk by launching a coordinated, country-wide cyberattack.

According to a James Derleth article in *Military Review*, another threat comes from "information warfare," a complex process that makes extensive use of digital communications. Derleth, the senior training advisor at a Europe-based training center, states:

> *Russian military leaders believe that a conflict's decisive battles are in the information domain and that information operations in the early phases are more decisive than later conventional warfare ... IW can create or leverage local military and political support, discredit leadership, slow decision-making, nurture dissent, shape public opinion, foster or manipulate local sources of instability, and mobilize local populations against foreign forces.*

Derleth explains that Russia used information warfare to annex part of Crimea and Estonia and asserts that the U.S. Army is not yet prepared for broad-scale information warfare.

## 10 kinds of weapons

The male drive for power has become a hydra-headed monster that subjects humanity not just to new kinds of weapons, but to new kinds of war, along with the old. As if that were not enough, the lethality of all these ways of war may soon be aggravated by artificial intelligence. Here are 10 kinds of weapons, some new, some old:

- Conventional explosives delivered by bullets, bombs, artillery shells, and rockets.
- Nuclear and thermonuclear, radiological, chemical, and biological – delivered by missiles or drones.

- Advanced hypersonic missiles.

- Smart drones able to attack in coordinated groups.

- Cyber attacks that disrupt a country's electronic and communications infrastructure.

- Cyber weapons, used by one country to gain control of another's satellites.

- Information warfare, a broad variety of methods designed to confuse, divide, and mislead a country's residents.

- In development: advanced lasers powerful enough to kill on earth while loitering in orbit.

- The practice of martyrdom, which expands the potential location of enemy attacks, lowers their cost, and simplifies their execution.

- Automatic warfare guided by autonomous artificial intelligence – not strictly a weapon, but a force multiplier.

This new diversity of weapons and their augmented deliverability make war more practical and therefore more likely. For example: One country attacks another with a nonviolent kind of warfare. The cost is relatively low, and the risk of serious reprisal is also thought to be low. But leaders of the target country, fearing a paralyzing loss of their command-and-control ability, respond with basic conventional weapons on a small scale. The attacker answers with more destructive weapons or with conventional weapons on a larger scale. The scale and lethality of the conflict increase as other countries jump in with new kinds of weapons to defend the interests of the side they favor.

The March 17, 2023 issue of *The Week* quotes a German journalist as saying, "If Iran goes nuclear, then its enemies in the region – notably Saudi Arabia, and possibly Egypt and Turkey as well – will likely seek atomic arsenals of their own." A relatively minor conflict quickly becomes a major war involving several countries. Some observers think some form of this scenario is likely to occur.

In a 2020 *New York Times* article, former CIA operative Paul Kolbe emphasizes the threat of cyber warfare:

> *First, the United States should recognize that it has entered an age of perpetual cyber conflict. Unlike conventional wars, we cannot end this fight by withdrawing troops from the battlefield. For the indefinite future, our adversaries, large and small, will test our defenses, attack our networks and steal our information.*

How does it all end? The sobering fact is that we don't know. There is no standing structure or agreement with the power to turn back the hands of the Doomsday Clock. Near the end of this book, we will explore how changing the gender profile of government could help to reduce the number of wars.

## From spear to missile

It should come as no surprise that brilliant people have been employed to push weapons development to extreme lengths. Throughout history – and indeed, before history – a lethal weapon was never lethal enough: "If you have a club, I need a spear; if you have a spear, I need a cannon; if you have a cannon I need a missile. Otherwise, I will be less of a man." And so it goes, on and on.

Our love affair with weapons probably started with a sturdy branch or antelope thigh bone. The next step was wooden spears, which enabled our ancestors to bring down more animals than was possible with just a club. The following step was a big one – attaching stone points to wooden spears. The superior penetrating power of sharpened stone enabled men to kill larger animals, from a greater distance. After that, things started to move toward higher tech, in steps reported by newscientist .com in 2009:

> *The atlatl, a way of using a short stick to add leverage to a spear, improved the range and accuracy of spear-throwing. It was a major advance in the effectiveness of human weapons, but it didn't satisfy our relentless hunger for more lethal weapons.*

The next step occurred when some genius thought of attaching a cord to both ends of a flexible length of wood and thereby invented the bow and arrow. The earliest potential examples date from 70,000 years ago, but the weapon was not widely used until about 3,000 BC. Now we could kill from an even greater distance, with higher accuracy.

About 3,300 years ago, with the advent of the Bronze Age, we took to using daggers and swords. Now one man could kill another, even without a spear or arrow.

About the same time, we started using the stirrup, which enabled men to lean over the side of a moving horse and slash at their opponents below. Cavalry was born, adding speed and mobility to the lethal power of an attacking unit.

About a thousand years ago, the Chinese developed gunpowder, which greatly increased the pace of weapons development. Freed from the limitations of muscle power, we could now invent (and use) a never-ending series of killing tools based on the explosive power of chemical compounds. The first gunpowder weapon was not a gun, but a cannon, since it was easier to make a large weapon than a small one. Now it was possible to kill more than one person at a time.

Turning to the American Revolution, the most common weapon was the smoothbore flintlock musket, which was portable and relatively easy to manufacture. By the time of the American Civil War, we were using rifled muskets, breech loaders, a few repeating weapons, and many cannons. Rifling – spiral grooves inside a gun barrel – increased the accuracy and range of portable infantry weapons. As mentioned in our national anthem, rockets were also used, along with swords, pistols, and the Gatling gun, which could fire 350 to 400 rounds per minute.

Skip to World War II. The machine gun had already changed the nature of land battles, and now we graduated to even more lethal weapons: Bombs delivered by aircraft. Cannons mounted on tanks. Long-range rockets like the German V2. And, of course, the atom bomb, which the U.S. used to devastate Hiroshima and Nagasaki in 1945. According to sciencealert.com, the Hiroshima bomb killed between 90,000 and 146,000 people with a force equal to 15,000 tons of TNT.

The Soviet Union tested its own atomic bomb in 1949. The United States could have decided that parity was probably the best we could expect against the Soviet Union. Instead, the U.S. upped the ante by exploding the world's first hydrogen bomb in 1952. It had a force equal to 10.4 million tons of TNT, making it roughly 700 times more powerful than the Hiroshima bomb. It vaporized an entire island and left a crater 164 feet deep. Now we had an ultimate weapon. Anyone who attacked the U.S. had to expect their country would be turned into a radioactive wasteland. Was it time to stop focusing on weapons and devote most of those resources to improving people's lives? Apparently not.

In 1954 the United States tested more powerful devices. One, called Castle Bravo, was designed to yield just six megatons (six million tons of TNT). Instead, the explosion produced a force of 15 megatons and irradiated distant Marshall Island residents who were supposed to be safe. Not to be outdone, the Soviet Union exploded several 20-megaton bombs in 1961 and 1962. A 1961 Soviet bomb produced a blast of 50 megatons. Called Tsar Bomba, it was the largest man-made explosion in history, 3,333 times more powerful than the Hiroshima bomb, according to an August 2020, *New York Times* article. It stands as the final result of the U.S. and Russian drive to produce ever-more-powerful nuclear weapons to "assure national security."

Meanwhile, the Soviet Union launched the first intercontinental ballistic missile (ICBM) in 1957. Now men could put an atomic bomb inside an ICBM and send it anywhere in the world. Was that enough kill-power? Not a chance. The next step was a mouthful: multiple independently targetable reentry vehicles (MIRVS), which enable one warhead to carry several missiles, each of which can hit a different target.

Skip to today, when the American helicopter-mounted Vulcan gun can fire 7,000 rounds per minute – enough to destroy an entire village in a single burst. Still not lethal enough? There's also a weapon called Metal Storm, which can fire 1.6 million bullets a minute. Coming soon, a broad variety of advanced weapons: chemical, biological, cyber, directed-energy laser, ultra-long-range artillery, electronic rail guns firing projectiles at 5,000 miles per hour, weaponized drones that select targets on their own, and, in the planning stage, kinetic bombardment satellites that fire at earth targets from orbit. That's just a partial list and includes only those the public knows about. Perhaps most troubling are the hypersonic missiles mentioned earlier. Unlike ballistic missiles, which tend to follow a predictable course, hypersonic missiles are maneuverable in flight, which would make them extremely difficult to track. As of 2020, at least five countries had them. An August 8, 2019, *New York Times* article reports,

*After the recent death of the treaty covering intermediate range missiles, a new arms race appears to be taking shape, drawing in more players, more money, and more weapons at a time of increased global instability and anxiety about nuclear proliferation.*

It's revealing to compare U.S. military spending with that of our main competitors. For fiscal year 2022, the U.S. military budget was $877 billion. That sum is 10 times the amount spent by Russia and three times the amount spent by China. (Axios, 4/24/23). For various reasons, it can be argued that the United States needs to have an edge in military prowess over its main competitors. But shouldn't we ask if some of that money would be better spent in other ways?

A lot of weapons development is proudly announced in news and articles. But hidden behind the pictures, we can see our closest primate relatives slipping through the jungle, intent on killing their male neighbors and taking their resources. Men's insatiable drive for ever-greater destructive power races on, with no end in sight.

## A zero-sum game?

Earlier in this section, the phrase, "Not to be outdone," appeared. It was not just a connective – it's probable that the male drive to assert dominance is a major motivator in the history of weapons. As discussed, men prize power. One reason may be that more power means more sex, more food, more status. In all of our history, so far as we know, almost every major new weapon has been invented by, developed by, and used by a single sex, the unspoken war cry being, "My sperm, not yours." We love our weapons, and we love to use them. How else to explain our obsession with ever-more-lethal weapons that make little sense in terms of making a nation more secure? Whenever a country steps up its kill-power to get ahead of its enemies, its enemies are driven to develop effective counter measures or even more powerful weapons. We saw how well that works in the history of nuclear testing by the United States and Russia.

Another example is what happened between India and Pakistan. The residents of India are known to be very bright people. Yet in 1974 the government saw fit to explode an atomic bomb and followed up in 1998 with five more explosions. This they did presumably to get ahead of their arch-enemy, Pakistan. Not to be outdone, Pakistan exploded five bombs of its own, shortly after India began its 1998 tests. India could have presented the 1974 explosion as a mere scientific experiment and stopped testing. But no, it had to explode more bombs and goad Pakistan to launch and accelerate its own nuclear test program. After spending a substantial part of a beefed-up military budget, all that India achieved was parity at a more dangerous (and expensive) level. The history of arms races indicates that proliferating advanced weapons is all too often a zero-sum game.

None of this is meant as support for a national policy of pacifism. Many countries have a cabal of aggressive men intent on pushing for dominance of one kind or another. They represent an ongoing threat that must be defended against in some form. At the same time, it's reasonable to expect that a much higher percentage of women with political power would insist on credible answers to vital questions, such as: "Will this advanced weapon make a significant improvement in our country's security? If so, will it do that cost-effectively? Or is there a better way to improve our security?" We can also expect that women would pay closer attention to Robert Gates's prescription for using a broad range of means and methods to achieve a nation's goals, not just military action.

## A different kind of threat

When Russia and the United States faced off in the Cold War, power in both countries lay in the hands of sane people, who, despite whatever faults they may have had, did not want to see their own cities destroyed, and, at bottom, had no desire to destroy the cities of another country. But now there are players on the scene who have no such reservations: men in terrorist organizations like ISIS and Al-Shabaab. It should be recognized that the concept of martyrdom is itself a powerful weapon. It means that terrorists see little or no restrictions in taking lethal action – no innocent people to spare, no hospitals to save, no weddings or funerals to avoid, not even concern for their own operatives. It doesn't seem to matter if they kill people who profess the same religious doctrines as the terrorists themselves; the victims are seen as martyrs to the cause. People with a different faith are "infidels" and deserve to be killed. It is "God's will." *The New York Times* once reported a revealing comment by a terrorist leader. Paraphrasing, it was, "You Westerners love life; we love death."

*Note:* Of course, these attitudes are not endemic to any particular religion. Instead, they come from a small group of people with perverse interpretations of basically benign religions. Peaceful people who follow those religions should not – must not – be tarred with the same brush as terrorists.

Another example of terrorist ideology is reported by philosopher and author David L. Smith. In his book, *The Most Dangerous Animal*, Smith describes comments by Ayatollah Khomeini, leader of the Iranian revolution, during a speech in honor of Muhammad's birthday:

> *War is a blessing to the world and for all nations. It is God*
> *who incites men to fight and to kill ... Thanks to God, our*

*young people are now, to the limits of their means, putting
God's commandments into action. They know that to kill
the unbelievers is one of man's greatest missions.*

In 1979, Khomeini was named Iran's political and religious leader for life and held that position until he died a decade later.

What these attitudes mean is that terrorists feel morally justified to kill thousands of unbelievers, and if this action brings retaliation that kills thousands of believers, that result is entirely acceptable. As martyrs, they will all be welcomed into heaven. A *New York Times* 2020 article by Mujib Mashal quotes a terrorist leader as saying, "We see this fight as worship. So if a brother is killed, the second brother won't disappoint God's wish – he'll step into his brother's shoes." The world has already seen the effect of these attitudes at work, most notably in the September 11 attacks on the United States, which killed people of many faiths.

It's easy to imagine a scenario in which stateless terrorists could acquire and employ weapons of mass destruction – nuclear, chemical, or biological: One or two disgruntled or deranged scientists provide the technology in return for millions of dollars. (History indicates terrorists can obtain substantial funding.) The terrorists hire a few engineers, chemists, or biologists to provide the necessary apparatus, which they use to attack Israel, hoping to gain the support of other nations. The agent of attack could be people who want to be martyrs, or it could be weaponized drones, like those used to attack the Saudi oil facilities. Israel then assumes the attack came from, or was supported by, its archenemy, Iran, and attacks an Iranian city. Russia decides it must show support for its ally and takes military action against Israel. This prompts the United States to act against Russia, and we are off and running to a global war.

Notice that the situation is all the more unpredictable, because the Cold War concept of Mutually Assured Destruction has lost much of its force: Even a major attack by stateless terrorists leaves little or nothing to retaliate against – no capital city to destroy, no standing army to attack, no production facilities to decimate. All the attacked nation can do is attack a lone terrorist or bomb a suspected concentration of terrorists, which is likely to kill many innocent civilians. The attacked nation knows these facts, but nevertheless feels intense pressure to lash out at somebody, somewhere.

Unfortunately, the potential failure of Mutually Assured Destruction as a strategy for preventing war is not confined to terrorist groups. In the December 4-2020, issue of the *Bulletin of the Atomic Scientists*, Peter Hatemi and Rose McDermott point out three critical flaws in relying on that strategy: potential lack of rationality in leaders, atomic weapons in the hands of non-state groups, and faulty or misunderstood signaling of intentions and resolve.

## Another threat

Visit the mind of an intercontinental missile as it streaks toward its target. The missile has been instructed to hit a terrorist group unless there is a wedding in progress. As it nears impact, the missile sees a much smaller crowd than its software expects to see at a wedding. Based on that anomaly, it decides to strike, despite its prior instructions. After impact, the missile senders have differing opinions. Some think the missile was right; others, that the wedding party had mostly dispersed, but was still in progress. It doesn't matter. The missile made the call.

What's going on here? A missile is a thing, and things don't make subjective judgments or decisions. Yes, but that was before the arrival of autonomous artificial intelligence (AI for short). Now pundits and experts tell us the civilized world has undergone a major change. A

May 30, 2023, *New York Times* article reported that a group of more than 350 executives, researchers, and engineers signed this one-sentence statement: "Mitigating the risk of extinction from A.I. should be a global priority alongside other societal-scale risks, such as pandemics and nuclear war."

Reinforcing that warning, pundits tell us AI could be used to freeze financial markets, shut down power grids, and stop water supplies, propelling the global economy into chaos. AI could even trick a nuclear nation into thinking a hostile nation had launched a missile attack against it, which could incite a nuclear war.

Other authorities think the doom-saying is over-blown. Artificial intelligence models are powerful, but not inherently dangerous to the world, they say. *The Economist* on April 19, 2023, published this summary: "Their text-generating prowess allows them to act as general-purpose reasoning engines. They can follow instructions, generate plans, and issue commands for other systems to carry out." In short, AI can be seen as just another powerful tool in the human quest to master our environment. However comforting that thought may be, it's worth noticing that some forms of AI deliver their product at incredible speed. Want 500 words on the history of missiles? It's yours in minutes.

**Automatic warfare?**

This combination of power and speed suggests that some configurations of AI could be used to automate military operations. In a world where one nation (or terrorist group) can overwhelm another in minutes, there would be a premium on super-fast response – and an automatic response would be the fastest. It would eliminate the remote but possible situation where an unforeseen glitch – electronic or procedural – delays human response.

As a national leader, you would have to prepare for the possibility that an enemy will marshal his forces in a way that threatens your country. You and your advisors might decide that the best way to protect your country – or to implement aggressive objectives – is to arrange for instant automatic military action, a move that would also act as a deterrent.

You would have your generals employ autonomous AI systems in selected military units to implement your decision. You would be told that any extended military operation must include troops on the ground. In a world of automatic warfare, this might take the form of special units traveling around the clock in vertical-take-off-and-landing aircraft. Working in shifts, such troops would be equipped with four kinds of clothing and gear, enabling them to deal with land destroyed by radiation, lethal gas, pathogens, or conventional arms. An AI system would tell such troops what to wear and when to deploy.

The advantages of this kind of maximal preparation might be viewed as so strong that many nations would arrange for an automatic military response. Is that far-fetched? It's easy to think so – easy to believe this situation would never occur because national leaders are responsible people who realize that automatic warfare would greatly increase the chances of nations going to war. As the *Bulletin of the Atomic Scientists* has shown, the possibility of catastrophic war is already prohibitively high. Widespread automatic warfare arrangements would have a multiplier effect, expanding the current risk of war and accelerating the pace of events. The result could be disastrous for humanity. With all that in mind, even ambitious governments would presumably not arrange for automatic warfare.

That kind of thinking is calming, but it tends to overlook certain realities about the history of people doing the unthinkable: A large and generally responsible nation saw fit to explode an atomic bomb in two enemy cities, when a similar effect might have been achieved by exploding the bombs outside the cities, close enough for residents to see their destructive power. Advanced nations have all too often slaughtered millions of civilians for no apparent gain in power or resources. In 2022 a major military power (Russia) invaded a country that was no threat of any kind. The list of gratuitous lethal aggressions is long and ageless.

Under it all, and perhaps driving it all, is the inborn propensity of human males – especially male leaders – to assert their dominance through violence and war: "I must be seen as strong and decisive, willing and able to assert our power at any time."

All of which tends to support the position of those who call for a major increase in the political power of a sex not known for an obsessive interest in dominance and power at any cost.

## A destructive tilt

Whatever the main cause of male aggression, it's clear that our drive for dominance can tilt ideologies toward war, as in these examples of dominance drive at work:

- Japan in the late 1930s and early 1940s: "We must fulfill our destiny to control the South Pacific and parts of China."
- Germany in the 1930s and early 1940s: "We must fulfill our destiny to control all of Europe."
- The United States in the mid-1800s: "We must fulfill our destiny to invade and occupy all the land west of the Mississippi."
- The Soviet Union, starting in the mid-1940s: "We must fulfill our destiny to export Soviet-style communism to the rest of the world."

Clearly, the male drive for dominance, as expressed by armed invasions, can lead to lethal combat on a massive scale, despite the deterrents of death and destruction.

## Ominous history

Despite all these threats, there may be some who will label them as "theoretically possible, but so improbable that we needn't take action to avoid them." University of California professor Jared Diamond has something to say about that in his book, *Collapse: How Societies Choose to Fail or Succeed*. Diamond points out:

> *That is, Tainter's reasoning [archaeologist Joseph Tainter] suggested to him that complex societies are not likely to allow themselves to collapse through failure to manage their environmental resources. Yet it is clear from all the cases discussed in this book that precisely such a failure has happened repeatedly.*

"Such a failure has happened repeatedly" suggests that humans – or at least human men – resist learning from history, and will continue to make war while destroying the environment. This is a major point and deserves to be taken seriously by voters and world leaders.

Diamond believes the cause is defective group decisions and cites four factors that contribute to such failures: (1) not anticipating a problem; (2) not perceiving a problem when it arrives; (3) not making a concerted effort to solve the problem; (4) not being able to solve the problem, even with effort. *Collapse* was published in 2005, so Diamond must have written it in prior years, a long time before the appearance of hypersonic missiles and bomb-laden smart drones. It was also before global warming became a *cause célèbre*, but many years after scientists reported that human-caused warming was affecting global climate. So once again, the kind of situation that Diamond reports in *Collapse* is upon us: Intelligent and informed officials can fail – and often have failed – to correct dangerous conditions. Will we wake up in time? The answer is not clear.

## Summary

Humanity now faces an unprecedented coalescence of at least three major factors:

*First,* the global military situation is controlled by men, many of whom – especially those in power – have an abiding propensity for combat and war. It almost doesn't matter whether men have this propensity by biological inheritance, culture, or both. What matters is that they have it, as demonstrated by our history of almost continual armed combat for thousands of years.

*Second,* the world is endangered by the proliferation of new kinds of weapons with enhanced deliverability – robotically controlled drones, maneuverable hypersonic missiles, various kinds of cyberattacks on earth and in space, and the possibility of automatic warfare directed by autonomous artificial intelligence – all of which come in a context of human-caused global warming, with a scarcity of fresh water, devastating fires, floods, and crop failures. That a stateless group of rebels had the means to crush Saudi oil production indicates the threat to global security has been raised to a new level. Powered by advances in artificial intelligence, component miniaturization, and declining production costs, drone technology has become a formidable weapon of war, made even more threatening by its apparent availability to terrorists seeking to advance a partly religious agenda.

*Third,* there were always terrorists, but now we have terrorists who use the concept of religiously motivated martyrdom to justify mass murder. They divide the world into believers and unbelievers, whom they refer to as "infidels." According to the terrorist code, infidels deserve to be killed. If believers are also killed by a terrorist attack, or by retaliation for such an attack, that is no deterrent, because the believers become martyrs.

An unwelcome addition to the foreign terrorist threat is the domestic threat, exemplified by the January 6, 2021, violent attack on Congress, which included armed organized groups with a record of prior violence.

The combined effect of all these factors is to establish a higher level of threat to long-term peace and security than the world has previously seen. As mentioned earlier, the atomic scientists' Doomsday Clock in 2023 was moved to 90 seconds to midnight – the closest to global catastrophe it has ever been. Men's insatiable drive for ever-greater destructive power races on, with no end in sight.

# 5

## Two Invasive Species – Part I

It's helpful to understand the behavior of human males by studying the behavior of chimpanzee males, with which we share more than 95 to 98 percent of our maternal DNA. The key to this understanding was supplied by Harvard professor Richard Wrangham and author Dale Peterson in their book, *Demonic Males*. They reported that only chimpanzees and humans share this pair of characteristics:

> Both live in patriarchal, male-bonded communities.
> Both engage in male-driven, lethal intergroup raiding.

This singular finding lays the foundation for observing several compelling parallels between chimpanzees and humans. We'll look at chimpanzees first.

### 1. Same ancestor

Six or seven million years ago, the descendants of a primitive ape split into two lines. One led to the genus *Pan*, which now includes chimpanzees (*Pan troglodytes*) and bonobos (*Pan paniscus*); the other led to various species of *Homo,* including us. Who was that primitive ape? For a long time, paleontologists had no idea. That changed in 2014 when fossil hunters in Kenya found a 13-million-year-old skull of an infant which they named "Alesi." In 2017, a *LiveScience/Scientific American* article describes how the scientists decided the skull resembles that of a gibbon and belongs to a new species,

*Nyanzipithecus alesi.* They concluded that this species was close to the origin of living apes and humans and that it originated in Africa. That conclusion fits with a massive amount of fossil evidence and DNA studies indicating that *Homo sapiens* became a species somewhere in Africa.

## 2. Similar DNA

First, a technical point: Most of the DNA we share with chimpanzees and bonobos is not located in the nucleus of our cells; it's in the mitochondria, which are specialized cellular structures that convert energy from food into a form that cells can use. Mitochondrial DNA (mtDNA) is inherited only from mothers. When molecular biologists use it to trace ancestry, they learn about a person's mother and a long line of grandmothers, not about the father and grandfathers.

You may have seen articles stating that humans share 98 or even 99 percent of their DNA with chimpanzees. Those figures, which refer only to mtDNA, have come to be regarded as established fact. But are they accurate? The answer may depend on which parts of the two genomes (a complete set of genes) you count. Some molecular biologists have lowered the figure to 95 percent, but even that may be too high. An *Evolution News* article reports that geneticist Richard Buggs puts the figure at only 82 to 84 percent. Why the big difference?

A concise attempt to explain comes from Tom Ward at futurism.com: "In short, we are 99% chimp, but only if you exclude 25% of our genetic material from the study and 18% of theirs." Opposing this view, other scientists believe it's correct and valid to count most of the genome parts that Buggs excludes. For our purpose, the exact figure doesn't matter. The fact remains that we share a high percentage of our mtDNA with chimpanzees and their cousins, bonobos.

### 3. Similar behavior

A 2019 book by Michael Thomasello, *Becoming Human*, lists eight behavioral similarities between us and chimpanzees. Summarizing his remarks:

There is recent research demonstrating that at least some great apes engage in humanlike activities: Make and use tools; communicate intentionally; have a kind of 'theory of mind;' acquire some behaviors via social learning (leading to 'culture'); hunt together in groups; have 'friends' with whom they preferentially groom and form alliances; actively help others; and evaluate and reciprocate one another's social actions.

Other similarities to human behavior include empathy and compassion, but also bullying, rape, infanticide, and the use of infantry squad tactics to invade a neighboring territory. To prepare his list, Thomasello may have drawn on the work of Jane Goodall and many other researchers, including Sylvia Amsler, John Mitani (University of Michigan), Craig Stanford (University of Southern California), Jill Pruetz (Iowa State University), Frans de Waal (Emory University), David Watts (Yale University), and Richard Wrangham (Harvard University), to name just a few of the scientists who have given us a rich body of work detailing the behavioral similarities between chimpanzees and humans.

### 4. Same drive to hunt prey animals

For several decades, the case for cooperative killing by chimpanzees was made by scientists who studied wild chimpanzees in African rainforests. The first to report this kind of behavior was Jane Goodall, in the seventies. Her early reports, beginning in 1960, had often described chimpanzees as loving, compassionate animals with many behaviors and patterns of socialization similar to ours. She also reported – and this upset prevailing beliefs – that chimpanzees made and used tools to fish for termites concealed in mounds. It meant that humans could no longer be defined as "the tool-using animal."

Later, Goodall reported that male Gombe chimpanzees cooperated in killing prey animals. Her descriptions of hunting tactics were precise:

> *While some members of a group would block escape routes, one chimpanzee would climb to where a colobus monkey was trying to hide and kill it. Afterwards, the killer shared the meat in response to begging behavior from females and allies, but did not share with rivals.*

Notice that this observation supports primatologist Craig Stanford's belief that sharing meat, not just eating it, was a vital factor in fostering growth of the human brain. Other researchers have also seen male chimpanzees hunt and kill monkeys. A 2015 bbc.com article reports on observations by John Mitani and David Watts, who spent many years studying a group of chimpanzees in Uganda, Africa:

> *Between 1995 and 2014, Watts and Mitani observed 556 hunts, 356 of which targeted red colobus. These hunts were very successful: 912 colobus were killed, an average of 3 per hunt. It has been clear for several years that the colobus population has declined as a result. A 2011 study found that the population fell by 89% between 1975 and 2007. In the late 1990s the chimps were killing up to half the red colobus population every year.*

Chimpanzees don't limit their hunting to monkeys. They also hunt small antelopes and piglets, both of which they kill by biting and by grabbing the animal in the same way we might, then repeatedly slamming it to the ground. They use a more sophisticated approach to hunt small primates called bush babies. A 2007 Iowa State University news release reports that researchers led by Jill Pruetz documented wild chimpanzees on an African savanna using weapons to hunt bush babies. The attackers, mostly female, turn a branch into a short spear by breaking off one or two ends and using their teeth to sharpen one end. Then they repeatedly jab the stick into likely tree hollows where the petite bush babies often sleep. This account tells us some chimpanzees don't just use weapons; they make them, using a several-step process.

That finding brings a new dimension to our understanding of chimpanzee and human lethal aggression: Those who believe that war is learned behavior see fights between males for mating rights as a normal part of animal life—natural selection at work—while war is seen as culturally driven aberrant behavior, forcing humans away from their naturally peaceful nature. Mitani, Watts, and Amsler, along with Wrangham and Stanford, have shown that territorial aggression as conducted by chimpanzees is also natural selection at work. Successful group aggressors gain more territory, get more food, and mate with more females. The reasons to fight for territory are at least as powerful as those that drive the fight for mating rights.

Many other scholars and authors who have studied the data tend to agree with this conclusion. They include Steven LeBlanc, Harvard archaeologist, Thomas Hayden, author; Keith Otterbein, State University of New York at Buffalo; Malcolm Potts, at the University of California; Kirkpatrick Sale, author; and David Livingston Smith, at the University of New England.

As mentioned, there are still some scholars and authors who don't agree that men have an innate proclivity for cooperative killing. They believe it is learned behavior, a product of culture, missing the point that culture comes from biology interacting with an environment. We will review their opinions in a subsequent chapter.

## 5. Similar drive to invade and kill

Since male chimpanzees kill prey cooperatively, it should come as no surprise that they, like us, also cooperate to kill their own kind. In 1974, Goodall and her associates began to report on behavior that was called "The Four-Year War." One group of chimpanzees conducted raids into a neighboring territory to kill its male residents. The attackers took no prisoners. The war continued until all the male members of the invaded territory were killed.

That last point looms large: The invading chimpanzees did not stop killing until every last male neighbor was beaten and bitten to death. Events like these may provide a clue to the origins of human genocide.

In her 2000 book, *Through a Window: My Thirty Years with the Chimpanzees of Gombe*, Goodall states:

> *Thus it is fascinating as well as shocking to learn that chim-*
> *panzees show hostile, aggressive territorial behavior that is*
> *not unlike certain forms of primitive human warfare ...*
> *Chimpanzees also show differential behaviour towards*
> *group and non-group members...non-community members*
> *may be attacked so fiercely that they die from their wounds.*
> *And this is not simple "fear of strangers"-- members of the*
> *Kahanma community were familiar to the Kasekela aggres-*
> *sors, yet they were attacked brutally.*

All of which may have moved Goodall to remark, in a review comment for *Sex and War*, by Malcolm Potts and Thomas Hayden, "I do believe we have inherited aggressive tendencies from our ancient primate past – but also traits of compassion and altruism that we observe in chimpanzees as well."

The Goodall group was not alone in witnessing chimpanzees' war-like territorial behavior. In 2010, John Mitani, David Watts, and Sylvia Amsler published the article, "Lethal intergroup aggression leads to territorial expansion in wild chimpanzees," in *Current Biology*. They begin by explaining the background of the current report: There were two prior instances of aggression that resulted in territorial gains, but each was alleged to have a technical flaw; another study was necessary. This excerpt is taken from a summary of their 2010 report:

> *Chimpanzees make lethal coalitionary attacks on members of other groups. This behavior generates considerable attention because it resembles lethal intergroup raiding in humans ... Here we present data collected over 10 years from an unusually large chimpanzee community at Ngogo, Kibale National Park, Uganda. During this time, we observed the Ngogo chimpanzees kill or fatally wound 18 individuals from other groups; we inferred three additional cases of lethal group aggression based on circumstantial evidence. A causal link between lethal intergroup aggression and territorial expansion can be made now that the Ngogo chimpanzees use the area once occupied by some of their victims.*

The authors explain how militant chimpanzees invade a neighboring territory: Normally noisy, they move silently in single file, exactly as human infantry often does in hostile territory. The apparent objective is to surprise individuals or a small group and then attack. Invaders typically retreat if they encounter a group of chimpanzees obviously larger than the invading group.

It would be legitimate to ask if these sorties are deliberate attacks. Couldn't they just be accidental incursions by animals looking for new sources of food? Researchers who have spent years studying wild chimpanzees are clear in their response: Invading chimpanzees deliberately use hostile and violent means to gain new territory: A 2006 article by David Watts and others in the *American Journal of Primatology* explains the motivation:

> *The best-supported adaptive explanation for such behavior is that fission-fusion sociality allows opportunities for low-cost attacks that, when successful, enhance the food supply for members of the attackers' community, improve survivorship, and increase female fertility.*

Notice the high importance of those three rewards for the male invaders: More food, obtained by access to more fruit, as well as more prey animals. Longer life, enabled by more consistent, high-quality nutrition. And a higher rate of reproduction, enabled by access to more females. In short, invading chimpanzees stand to gain substantial rewards, similar to those of invading humans.

As mentioned earlier, both chimpanzees and humans live in patriarchal, male-bonded communities, and both engage in male-driven, lethal, intergroup raiding. Only these two species display both of these traits. Wolves attack their own kind, sometimes in raids. But they do not form patriarchal, male-bonded communities, nor do lions and other big cats. Hyena packs are dominated by females. The exclusive similarities between male humans and male chimpanzees also indicate why it's more useful to compare behavior patterns between humans and chimpanzees than between humans and bonobos.

We've seen that chimpanzees and humans have the same ancestor, similar DNA, similar general behavior, the same drive to hunt prey, and a similar propensity to invade and kill: brothers under the skin.

# 6

## Two Invasive Species – Part II

Turning to humans, one of the most intriguing issues in the history of human origins is why Neanderthals went extinct a short time after *Homo sapiens* moved into Europe about 40,000 years ago. There are several theories. One is that Neanderthals lived in small, separated groups, which led to harmful inbreeding. Another theory is that they didn't go extinct – they were assimilated into us, especially in Europe. But the proportion of DNA inherited has been measured in Europeans at only one to two percent.

**Neanderthal extinction**

The most widely held theory is that we did a lot of things better than they did. Curiously, almost every authority uses the same word to describe that situation: "out-competed." We outcompeted them in hunting, because we had better weapons. We outcompeted them in dealing with the cold, because we could make better clothing and build better shelters. We outcompeted them in many activities, because our language and communication skills were superior to theirs.

Each of these theories has its adherents, but a case can be made for another possibility. Neanderthals thrived in Europe for about 350,000 years before we showed up. That's longer than we're known to exist as a species. We were better at some things than the Neanderthals, but Europe is a big place; it must have had a plentiful supply of fresh water, edible plants, and prey animals. Also, the Neanderthals had certain advantages over us: Their thick torsos were well-adapted to the cold. They knew how to make the most of their environment – knew the seasonal movements of prey animals, knew where to find edible plants, knew where to find the most habitable caves and drinkable water. We had to learn all that. To top it all off, the Neanderthal brain was typically bigger than ours. Their brain may not have had as many interconnections, but that point is unclear. So why did the Neanderthals go extinct?

One answer is the episode described in "First Battle," which began on the cover of this book. The dark-skins in this imagined encounter were Cro-Magnons, a group of *Homo sapiens* who originally came from east Africa (thus the dark skins), migrated into the Near East, then turned northwest into southern Europe 35,000 to 40,000 years ago. This was a fateful move for the Neanderthals. Within about 10,000 years after the Cro-Magnons arrived, most of the Neanderthals were gone from their former territories; remnants managed to survive in what are now Gibraltar and Spain, but eventually, the species went extinct.

It is profoundly unpopular to say they went extinct mainly (not entirely) because we killed them. Learned scientists don't want to darken their reputation by presenting humans as a bunch of killers. It's obviously more politic to say that we had more adaptive genes, so we just naturally outcompeted them, without malice or intent to harm.

But what are the objective referents of "outcompete?" Do they include killing a high percentage of all the prey animals in Europe, leaving little for the Neanderthals? Taking over nearly all the sources of clean fresh water? Occupying all the desirable caves? Doing all those things and more in nearly all of Europe? Obviously not. Some level of competition must have occurred, but it's hard to imagine how it could have been enough to drive a highly successful, big-brained species into extinction. Instead, it may be that the "outcompete" theory, while partly valid, is mainly a way to protect us from having to confront an ugly truth: The males of our species invade and kill. In *Where Does Violence Come From?* neuroscientist Bernhard Bogerts comments,

> *…early humans perished not only because of climatic or other adverse living conditions, but also because they were wiped out by superior races of the genus Homo in the course of rival group fights and ethnic conflicts [similar to] pogroms over many generations.*

Recent examples include the U.S. invading Iraq and Russia invading Ukraine. History tells us invading and killing are what the males of our "advanced" species have always done. Paradoxically, this murderous behavior has come amid countless acts of caring and compassion.

## Men, the aggressors

News reports tell us about men with political or economic power who abuse women simply to assert dominance, an attitude epitomized by the infamous, "Grab 'em by the pussy." Others (often the same men) seem to live by a related dictum: "Rules are for nobodies, not for me." It's a curious fact of human nature that such men can rise to high positions, largely on aggression, with little else to offer. At the extreme, some male leaders order or promote needless military actions that destroy cities and kill tens of thousands of people, including women and children. In a grotesque example of male aggression, they perpetrate heinous crimes on a massive scale just to seize or hold power.

Sometimes the aggressive drive masquerades as an evil ideology, as in Nazi Germany, Stalin's Soviet Union, and currently, in Vladimir

Putin's Russia. Men who perpetrate these ideologies often attract followers, especially males willing to slaughter tens of thousands of other people, because their victims have a different religion, belong to a different race, or because they're defined as enemies by their leader. There's nothing new about any of that, but in the last few decades, the media have put it in our face, forcing us to take a closer look at aggression. We'll start with definitions:

## Aggression

Aggression in this book means any action intended to inflict harm on another entity; it may or may not be driven by anger. In contrast, assertion is taking action to gain an advantage or prevent a loss, with no intent of harming anyone; typically, it does not require anger.

If I plant a tree on my property, I'm asserting a property right. If I attack a neighbor's tree, I'm committing an act of aggression.

There are instances where it's hard to tell the difference. An NFL linebacker probably doesn't mean to injure the other team's running back. Nevertheless, what he does to stop the running back can fairly be described as aggressive. Similarly, when a lion attacks a zebra, harming her prey is not her primary objective – she gets no emotional satisfaction from it. She's just trying to get dinner. But when you see the attack happen, it seems to require extreme aggression. In both cases, it can be argued that aggressive methods are used to accomplish a non-aggressive objective.

## Violence

Aggression often leads to violence. As used here, violence refers to physically harming a person or another animal and to damaging property. If a hunter shoots a pheasant, she has committed an act of violence. Killing and torture are the most extreme forms of violence. But malicious intent is not required. If a driver strikes a pedestrian, he has perpetrated a violent act, even if the driver had no intention of causing harm.

Broadening our view, we see that physical violence is not the only form of aggression. In many elementary and high schools in the U.S., it's all too common to see students intimidated and humiliated almost every day. It takes many forms, including verbal abuse, grabbing property, blocking an intended route. The threat of imminent intimidation hangs in the air like a foul-smelling fog. The form of aggression varies, but the intent is always the same – to assert dominance, to intimidate, to humiliate, often with a rush of feel-good dopamine. In the adult world of work, there is plenty of aggression, although it's usually cloaked in apparent civility.

In his latest book, *Human Nature*, David Berlinski makes a point about Harvard professor Steven Pinker's book, *The Better Angels of Our Nature*, which argues that violence in the world has declined. Berlinski comments:

> *The years between the end of the Second World War and 2010 or 2011, Pinker designates the long peace. It is a peace that encompassed the Chinese Communist revolution, the partition of India, the Great Leap Forward, the ignominious Cultural Revolution, the suppression of Tibet, the Korean War, the French and American wars of Indochinese succession, the Egypt-Yemen war, the Franco-Algerian wars, the Israeli-Arab wars, the genocidal Pol Pot regime, the grotesque and sterile Iranian revolution, the Iran-Iraq war, ethnic cleansings in Rwanda, Burundi, and the former Yugoslavia, the farcical Russian and American invasions of Afghanistan, the American invasion of Iraq, and various massacres, sub-continental famines, squalid civil insurrections, blood-lettings, throat-slittings, death squads,*

*theological infamies, and suicide bombings taking place*
*from Latin America to East Timor.*

With that single paragraph, Berlinski exposes the truth about global male aggression without venturing a single formal argument; he simply states history.

## Signs of aggression

Some signs of aggression are pervasive in society. It's no accident that companies are said to "battle" for market share; imposing tariffs creates a "trade war;" and those who make a highly profitable transaction are said to "make a killing." Our most popular athletic sport is football, which requires violent collisions. The Superbowl, in fact, brings more Americans together than any other event – over 100 million people focus on our most war-like sport.

Even that figure does not match the popularity of guns in the U.S., which has more firearms (450 million) than people (330 million). According to the Small Arms Survey of 2017, Americans own six times as many guns per person as do citizens of Germany and France – and 24 times as many as in the United Kingdom. The U.S. has the distinction of being the only advanced country where gunshots are the leading cause of death for children and teens, according to a March 29, 2023, report from CNN.

Americans seem to have a blind spot about the high density of gun ownership. Few would argue with an analogous point – that the higher the density of cows in a pasture, the greater your chances of stepping in cow dung. But voters can't seem to apply the same logic to the number of guns. Further, they miss the point that guns make it too easy to kill – move one finger one-half-inch, and a potentially lethal bullet is on its way. The ease of pulling a trigger is one reason an average of more than 300 U.S. children per year use a gun to kill or injure other children and adults, according to Everytown research.

Summarizing, it appears that some forms of aggression have decreased, especially interpersonal aggression in Europe, but not its overall frequency. Physical aggression – often with lethal intent – appears to be a fundamental aspect of male humans.

# 7

## What Made Us Human?

Humans are part of a genus called *Homo* and a species called *sapiens*, meaning wise. Starting about two million years ago, there were at least nine species of *Homo*, with some living at the same time as our direct ancestors. About 30,000 years ago, we became the only surviving *Homo* species, a topic we will explore in the pages ahead.

**Part 1: Why are we *Homo*, instead of *Pan*?**

The question arises, in part, because of our similarities to both members of the *Pan* genus (chimps and bonobos). As stated earlier, we share more than 95 percent of our maternal DNA with chimps and bonobos. Beyond that basic fact, we have the same distant ancestor; we are primates and omnivorous; we kill and eat prey animals; groups of male chimpanzees (but not bonobos) invade and kill chimps in neighboring territories; humans and *Pan* species behave in very similar ways; we both have the primate suite, which includes highly developed brains (though ours is much bigger), depth perception, and opposable thumbs. Given all those similarities, why aren't we *Pan something or other?*

The most obvious answer is that we are descended from relatively large-brained primates who could walk upright, and that freed our hands to make and use tools and weapons. These stone creations – scrapers, hand-axes, cutters, and spear points – enabled their users to gain access to a wide variety of food, including meat, which helped to enlarge our brain. Other explanations include the point that habitual upright posture was useful to see predators and prey over tall grass. It also offered some protection from the scorching African sun, compared with exposing the back. As with many other aspects of the human story, it was likely a case of many factors coming together.

## The importance of throwing

Once some of our ancestors were able to walk more or less as humans do, there is evidence that we used shoulders, arms, and parts of the primate suite to get meat. A 2013 article in *Nature* by Neil Roach at Harvard, and others (including Dan Lieberman), describes the method. From the abstract:

> *Here we use experimental studies of humans throwing projectiles to show that our throwing capabilities largely result from several derived anatomical features that enable elastic energy storage and release at the shoulder. These features first appear together approximately 2 million years ago in the species Homo erectus. Taking into consideration archaeological evidence suggesting that hunting activity intensified around this time, we conclude that selection for throwing as a means to hunt probably had an important role in the evolution of the genus Homo.*

## The arrival of grass

One way to think about human origins is to start far back in primate history. A few million years ago, there were at least two kinds of primates roaming the woodlands and savannahs of Africa, the robust *Paranthropus* and the slimmer *Australopithecus*. They looked somewhat like apes, but with one big difference: They walked upright, not some of the time, as chimpanzees do, but all the time. Their diet consisted of fruit, nuts, berries, and leaves. They did well for hundreds of thousands of years, and then something happened. In fact, it was something that had happened before and would happen again:

The earth slowly became much colder, which may have occurred because the planet tilted in a way that reduced the amount of solar radiation striking the earth. In northern areas, this caused an ice age. In Africa, it brought alternating periods of cool, dry weather and warm, wet weather. The net effect was to turn vast areas of forest into grass and broken woodland. The grass, in turn, brought massive herds of grass-eating animals, including wildebeest, zebras, and many kinds of antelope. With these hefty bundles of protein and other nutrients came the predators – big cats, hyenas, jackals, wild dogs, and eventually, us.

In a 2014 *Scientific American* article, Kate Wong points out that the availability of prey animals set the stage for a split between *Paranthropus* and *Australopithecus* that had major consequences: The robust *Paranthropus* line dealt with the loss of forest food by developing giant jaws and big teeth to chew the tough tubers and other plants found in semi-dry environments. The slimmer *Australopithecus* line turned to eating meat, which proved to be the more adaptive solution: Plant-eating *Paranthropus* went extinct, while meat-eating *Australopithecus* became an early ancestor of the deadliest predator on earth.

## Part 2: Why are we *sapiens*?

To explore this topic, we start with an idea advanced by Mark Strauss in a 2015 *National Geographic* article, "12 Theories of What Made Us Human, and Why They Are All Wrong." Strauss considers 12 different theories and shows why each may be flawed, or at best, insufficient to explain what made us human. He concludes that the cause was not any single factor, but rather some combination of all 12 theories.

This general idea can be refined down to what might be called the Quest for Meat – 10 sequential steps that were a necessary part of a typical hunt by our ancestors. (There is one more task, which helped protect the productivity of a group's hunting grounds.) We'll scan the steps first, then amplify:

1. Making tools and weapons

2. Planning and organizing hunts

3. Stalking and ambushing

4. Spearing to wound

5. Tracking and chasing

6. Spearing to kill

7. Sharing meat

8. Cooking meat and plants

9. Eating meat, especially by young children

10. Enjoying the sexual reward

In addition, engaging in combat with territorial enemies or competitors was probably part of the hunting lifestyle.

Bear in mind that these actions could not have produced our human ancestors directly. In general, a creature's actions cannot change its DNA. Instead, change comes from the natural selection that flows from these actions, aided, perhaps, by favorable mutations.

## THE QUEST FOR MEAT

### *1. Making tools and weapons*

According to a BBC article, a search for African fossils in 2011 began with a detour. Researchers led by Sonia Harmand, of Stony Brook University, were walking through the arid landscape near Lake Turkana, Kenya, when they took a turn from their intended route. They had not gone far when they noticed a large protruding rock with an interesting face. Examining it with a knowing eye, they saw it was a core from which flakes had been chipped. By the end of 2012, they had found about 150 hammers, anvils, cores, and sharp-edged flakes. More have been found since then.

That would be a welcome find for any archeological expedition. But the best was yet to come. The stone tools turned out to be 3.3 million years old, which was 700,000 years older than the oldest tools found anywhere else. The discovery was not only a remarkable find in itself – it was also disruptive to an accepted understanding: Prior to this discovery, paleontologists were fairly sure they knew the identity of the first toolmaker: It was probably *Homo habilis* (handyman), historically regarded as the first species in the *Homo* genus. But now there was a problem: The oldest known *habilis* fossils dated from 2.4 million years ago, far less than the age of the Lake Turkana tools. So if it was not *Homo habilis*, then who made these tools? The suspects include a couple of pre-Homo species, but no one knows for sure. The upshot is that scientists have had to revise their thinking about the capabilities of early hominins: Either they were smarter than has been generally realized, or it takes less intelligence than was previously thought to make stone tools.

It may mean something else, as well. If the Turkana toolmakers wanted sharp-edged flakes, it could mean they were using the tools to scrape meat off skin and bones. The meat might have been scavenged, or, given the typical unreliability of scavenging, it could also have come from animals these early toolmakers killed. If that behavior was preserved by natural selection, it would have set the stage for humans to do the same much later in primate evolution.

We come to the more recent art of making weapons. According to a 2013 *Archaeology* article, the known history of weapon-making starts with 210 stone tools about 460,000 years old, found at a site called Kathu Pan in South Africa. A study led by Jayne Wilkins, then at the University of Toronto, showed that 23 spear points were thinned at the base, apparently to make them easier to attach to a shaft. Researchers tested this idea by firing spears made from replica points into springbok carcasses. The replica points were damaged in ways similar to the damage in the original points. Scientists speculated the spears had been used by *Homo heidelbergensis*, a species generally regarded as directly ancestral to Neanderthals and humans.

Making stone tools and weapons was not just a way to get meat. The actions required to produce these objects also stimulated growth of the brain in several ways. A 2017 *Nature* article states:

> *Here we show that [ancient] tool production [starting 1.76 million years ago] requires the integration of visual, auditory, and sensorimotor information in the middle and superior temporal cortex, the guidance of a visual working memory ... and also higher-order action planning ... activating a brain network that is also involved in modern piano playing.*

Toolmaking helped to develop the human use of language, as suggested in the title of a 2012 scientific article: "Stone tools, language and the brain in human evolution." The article's authors report, "Long-standing speculations and more recent hypotheses propose a variety of possible evolutionary connections between language, gesture, and tool use. These arguments have received important new support from neuroscientific research." Many scientific papers have been published on the relation between toolmaking and the development of speech.

## 2. Planning and organizing hunts

Three hundred thousand years ago, it was no casual thing to plan and organize a hunt. Just having a bunch of people charge out of a cave and start throwing spears at the first animals they saw was not likely to be successful. This, they would have learned by experience.

The first issue may have been who would be allowed to join the hunting team. In most cases, it was probably young men. By the time a typical female was old enough to thrust or throw a spear with sufficient force, she was probably pregnant, nursing, or had a toddler to care for. So she had to be protected, not subjected to physical risk. But females could have, and probably did, hunt and kill small game. We can assume they also gathered edible fruit and plants, which were essential for survival; it's likely that many hunts for large animals were unsuccessful, and people still had to eat. Females may have provided more food for proto-humans and early humans than did the males.

The next issue might have been, who would lead the hunt? Most of the time, it was probably the alpha male; people were used to listening to him, so members of the hunting party would tend to do as he ordered. Other issues involved deciding where and when to hunt. Once close to a target herd, the alpha male may have tried to get the hunters to space themselves according to his idea of the optimal placement. Most of the men on the hunt had to be willing to follow the leader's orders. Leaders need followers.

In all of this, the ability to communicate with specific language sounds would have been extremely useful. This ability was probably reinforced by natural selection, as were the cognitive abilities required for planning and organizing.

Some scientists are skeptical about how much proto-humans and early humans hunted. They explain butchering marks on two-million-year-old fossilized bones as the result of scavenging: We took meat from animals killed by predatory quadrupeds, fire, or illness. But as Craig Stanford notes in *The Hunting Apes*, most predators also scavenge. With that point in mind, it's likely our ancestors did both. Many relevant scientists doubt that scavenging alone could have produced enough meat for long-term survival.

Wild predators have now, and must have had then, a sense of smell far more powerful than ours. Considering all the types of predators – big cats, smaller cats, hyenas, jackals, wild dogs, eagles, vultures – there had to have been a large number of meat-eaters occupying and patrolling the land where prey animals lived.

With their superior sense of smell, speed, and ubiquity, four-legged predators and meat-eating birds, including vultures, probably got to carcasses long before two-legged primates. Also, there is no evidence that they politely stepped aside when humans appeared at a kill. Imagine: You are one of several healthy adult males about 250,000 years ago, and you arrive at a kill to find several lions or a pack of hyenas already feeding on the carcass. Your job is to scare them away with a stone-tipped tree branch. How, exactly, do you do that, without becoming part of the predators' meal? The obvious answer is that you don't, even with the help of a small group of hunters just like you. What you do, instead, is hide in the grass until the four-legged predators have filled their bellies and walk off. Even then, you had to contend with the vultures and eagles, which are not easily dislodged. With all these points in mind, it's hard to escape the conclusion that hunting provided more meat to early hominins than did scavenging. Hunting practically guaranteed that the human killers would be first on the scene of a kill.

## 3. Stalking and ambushing

Once local prey animals learned that the unfamiliar bipeds were dangerous, we can assume they shied away or ran away when humans approached. Which means the hunters had to stalk or ambush to make a kill. To stalk, they had only to imitate the methods of the big cats: Use the tall grass for cover, get close, then charge. Ambushing requires more planning. The hunters had to anticipate where their prey would pass, then find a suitable place from which to attack. Typically, this may have meant sitting in a tree next to an animal path. Stalking and ambushing required planning for an imagined encounter.

## 4. Spearing to wound

In some accounts of proto-human hunting, it's assumed they took their spears, walked out onto the savannah, and killed a prey animal. It may not have been that simple.

A male reader might like to imagine himself as a human on the hunt 200,000 years ago. As mentioned, all you have is a semi-straight tree branch with a stone point stuck on one end. True, there are other hunters, similarly equipped. But a 300-pound wildebeest mother is charging straight at you – she's decided you're a threat to her calf. If you jump away, you know you'll get no piece of the meat if a kill is made. Worse, you won't get any sex, either. Human females were not stupid, and maybe some are watching from a distance. You know you have to stick your spear into that wildebeest mother. Men who did that (and lived) passed on more DNA to future generations than those who fled the threat.

It's important to note there's a high probability that the first strike from a spear wounded the animal, but did not kill it. It seems likely that spears with rough stone points often came out of the wound hole as animals reacted to the painful penetration. This left the prey free to run. A rush of epinephrine (adrenaline) and cortisol gave a wounded animal the strength to dart away, sometimes for a long distance. Even today, hunters with powerful rifles sometimes have to follow a blood trail to finally kill their prey. So a spear strike was not the end of a hunt; it was just the beginning.

## 5. Tracking and chasing

Following and catching up to a wounded animal was no walk in the park. Much of the blood from a wound might have stuck to the animal's coat, leaving only a faint trail with spots of blood far apart. How did we succeed? It's important to realize we did not leave the forests of Africa ready to hunt and kill animals on the savannahs. Our ancestors had to live for many thousands of years in open woodlands, where they evolved the specific abilities needed to pursue prey and throw spears. These abilities include:

- The suite of body structures required for running on two legs: Short toes. Long-arch feet. Forward-facing legs and hips. Strong hindquarters. Balanced head and upper torso.
- The heart and lungs needed for "endurance running," which can mean running or jogging for at least 30 minutes.
- The ability to sweat, which dissipates the heat built up by jogging or running long enough to catch a target animal. (Many prey animals, lacking the ability to sweat, need to stop running after some distance or die of overheating.)

These criteria are based on a 2004 seminal paper by David Bramble (University of Utah) and Daniel Lieberman, of Harvard University. Before the publication of their paper, most scientists had focused on the body structures for walking. Bramble and Lieberman showed that endurance running requires a different body structure from the one used for walking. They also showed how those structures may have played a major role in the evolution of *Homo sapiens*: Endurance running became necessary as our ancestors pursued animals for meat, and meat, in turn, stimulated the growth of our brain. In the authors' words:

> *One possibility is that [endurance running] played a role in helping hominids exploit protein-rich resources such as meat, marrow and brain, first evident in the archaeological record at approximately 2.6 [million years] ago, coincident with the first appearance of Homo.* (That first instance of *Homo* may have been *Homo habilis*, a proto-human.)

The authors also point out that no other primate is capable of endurance running. Considering the similarity between humans and other great apes, that is a striking fact. It underscores how the desire for meat helped to produce the unique body structure and brain of *Homo sapiens*.

The plot thickens when we compare our ancestors' preference for meat (and its result) with the findings of two other scientific studies. One comes in a 2019 *Scientific American* article by Herman Pontzer. The author points out that *Homo naledi,* an extinct species discovered in South Africa, persisted as a lineage "quite happily for more than a million years without the continued increase in brain size seen in other *Homo* species." This surprising fact immediately raises the question, "Did they eat meat?"

Ian Towle, of Liverpool John Moores University, gives us the answer in a 2017 article in theconversation.com, in which he reports that *Homo naledi's* teeth were severely chipped:

> *In H. naledi more than 40% are affected – which is very high ... The back teeth are the most fractured, with more than half having at least one chip and many having multiple small chips ... For now, all we can conclude for certain is that H. naledi consumed a diet significantly different from any other fossil hominin species yet studied, containing a bigger proportion of small, hard objects.*

We now have a likely solution to the puzzle: We know that *Homo naledi's* brain did not grow over millennia; we also know it's the only known *Homo* species whose teeth indicate it was not a meat-eater. Current knowledge indicates the brain of at least some meat-eating *Homo* species did grow. Taken together, these points suggest the *Homo sapiens* brain grew, precisely because our persistent quest for meat set off the 10-part process we are currently exploring.

### 6. Spearing to kill

Imagine the scene as our hunting ancestors finally caught up with a wounded animal that has stopped to cool down and catch its breath. Unless the animal was grievously wounded by the first spear strike, it is still dangerous. The wound may have been slight, and the animal now knows, beyond any doubt, that the bipeds are its enemies. At this point, a few of the hunters have to move in close enough to try for a fatal strike with their crude spears, an act that required considerable skill and aggression. They cooperate in attacking from different angles in hopes of confusing their prey. That may work as planned. Or it may not, and one or more of the hunters is killed or wounded.

Still, they had to try. They needed the meat, and often enough, they got it.

Which poses a question: Why were some proto-humans and early humans skillful at killing large animals with crude spears? And why, today, can some Little League pitchers throw a baseball accurately at 50 miles per hour, and some Major Leaguers, at 100 miles per hour? The question is especially pertinent when we consider the throwing ability of chimpanzees: Full-grown chimpanzee males are much stronger than adult human males, but the fastest they can throw is a feeble 20 miles per hour, with little accuracy.

The Neil Roach article quoted earlier provides an answer with the discovery of elastic energy storage in the shoulder. What the researchers learned is that our flexible waist and shoulders can be bent back to store energy and then suddenly released, creating an analog to pressing on a spring and then letting go. No other primate can do that. For our ancestors, there remained the issue of accuracy. That probably happened the same way it does today, only instead of throwing balls, they threw spears. We can imagine that when a young person was able to show sufficient power and skill with a spear, he (or perhaps, she) was allowed to join the team of hunters.

## 7. Sharing meat

Some anthropologists emphasize sharing meat as a vital factor in the evolution of *Homo sapiens*. One such is Craig Stanford, a professor of anthropology at the University of Southern California. Working with Jane Goodall, he has spent many years studying chimpanzees and gorillas in Africa and monkeys in South Asia. Here are a few relevant excerpts from his book, *The Hunting Apes:*

> *In this book I argue that the origins of human intelligence are linked to the acquisition of meat, especially through the*

> *cognitive capacities necessary for the strategic sharing of meat with fellow group members...the intellect required to be a clever, strategic, and mindful sharer of meat is the essential recipe that led to the expansion of the human brain.*

Addressing those who tend to downplay the importance of meat in human evolution, Stanford makes a useful distinction: It's likely that many hunts were unsuccessful, so meat may have supplied less than half the calories of some proto-humans and early humans. This would have made it a less valuable food source. But Stanford shows that it was nevertheless the food most *valued* by early hunters. Meat was central; everybody wanted it. This meant that whoever had meat – typically adult males – was in a controlling position. He could use it to lure females for sex, reward friends and deprive enemies, and form alliances that would strengthen his position in the group. Stanford states, "In the doling out of the meat we see signs of strategizing and politicking that go far beyond those seen in predatory behavior." Hunting called for planning, aggression, and certain physical abilities; sharing required conniving, which may demand a higher form of intelligence.

## 8. Cooking meat and plants

There is some divergence of opinion among scientists about the importance of cooking to the evolution of Homo sapiens. It's clear that control of fire and cooking brought substantial benefits to early humans and, perhaps, to some proto-humans. By the time *Homo erectus* appeared two million years ago, primates in the *Homo* line had lost their ability to climb trees with the same ease as chimpanzees and orangutans. That meant they had to live on the ground at night, as well as during the day, a lifestyle that exposed them to sudden attacks by prowling big cats and hyenas. Controlling fire would have helped to reduce that threat.

Fire provided warmth on cold nights and, importantly, provided a central point for socializing that fostered bonding among group members. Not so incidentally, it gave men a forum in which they could boast about how they braved the dangers of a hunt to bring back meat for the group.

Control of fire gave primates in the *Homo* line a way to improve tools and weapons: When heated, certain kinds of stones can be flaked and chipped to more exact shapes than unheated stones.

But the most influential use of fire was for cooking meat and other foods. In *Catching Fire*, Harvard Professor Richard Wrangham points out that cooking makes food safer, improves taste, reduces spoilage, and most importantly, "increases the amount of energy our bodies obtain from our food." This extra energy probably helped proto-humans and early humans run faster, throw harder, and think better. Scientists seem to agree on all that. What some don't agree on is when it started. Wrangham states,

> *Fossil evidence indicates that this dependence [on cooking] arose not just some tens of thousands of years ago, or even a few hundred thousand, but right back at the beginning of our time on Earth, at the start of the human evolution, by the habiline that became Homo erectus.*

That primate, possibly *Homo habilis*, first appeared about 2.5 million years ago. The timing is critical. If cooking didn't start until *Homo sapiens* appeared on the scene, it could not have had much to do with the evolution of humans. Charred bones, rocks, and ash from millions of years ago have been found, but many or most of these remains could have been produced by lightning strikes. Wrangham counters with other kinds of evidence:

> *The reduction in tooth size, the signs of increased energy availability in larger brains and bodies, the indication of smaller guts, and the ability to exploit new kinds of habitat all support the idea that cooking was responsible for the evolution of Homo erectus. Even the reduction in climbing ability fits the hypothesis that Homo erectus cooked ... This shift suggests that Homo erectus slept on the ground, a novel behavior that would have depended on their controlling fire to provide light to see predators and scare them away.*

Wrangham is not alone in emphasizing the importance and early use of cooking. In a livescience.com post in 2011, science writer Jennifer Welsh reported on a study published in *Proceedings of the National Academy of Sciences*. Early in Welsh's discussion of cooking, she points out there was an increase in absolute hominin brain size about two million years ago, but not to the same extent as the increase in body size. "Maybe, she says, "meat was not completely responsible – so what was?" She provides this answer:

> *Perhaps it was the shift from eating antelope steak tartare to barbecuing it. There are hints of human-controlled fires at a few sites dating back to between one and two million years ago in eastern and southern Africa, but the first solid evidence comes from a one-million-year-old site called Wonderwerk Cave in South Africa. In 2012, Francesco Berna, then of Boston University, and his colleagues reported bits of ash from burnt grass, leaves, brush, and bone fragments*

*inside the cave...It turns out that, using fossil kills to meas-*
*ure brain size, we see the biggest increase in brain size in*
*our evolutionary history right after we see the earliest evi-*
*dence for cooking in the archaeological record, so he*
*[Wrangham] may be on to something.*

Still-skeptical scientists say that if cooking occurred farther back in time, it may not have been widespread; the scarcity of ancient hearths tends to support that view. But maybe there's a role for common sense in this discussion. If so, the first point might be that lightning starts fires on savannahs today and probably did so two million years ago. Such fires would have roasted many carcasses of dead animals; the enticing odor would have carried to nearby groups of pre-humans or humans. They would have been able to get at least some of that meat and undoubtedly found it a lot better than raw meat. From there, it's no great leap to imagine that early hominins grabbed partially burn-ing sticks and carried them to wherever they congregated. Add some dry grass and wood, and you have a cooking fire. Not a big deal, and you don't need a big brain to do it.

But if all that is true, where are the paleo-hearths? The answer may involve population movements. We know there were relatively frequent changes in African climate – changes that would have prompted herds of animals to migrate, even as they do today. Given that early hominins were meat-eaters, it's likely they followed the movements of their favorite prey animals. If so, we should not expect to find many well-established hearth sites. After more than a million years, it's reasonable to expect that the physical evidence of sundry cooking fires would range from thin to non-existent.

### 9. *Eating meat, especially by young children*

Did eating meat make the difference between humans going extinct and achieving global dominance? It could have, if getting, sharing, and eating meat helped to give our ancestors a bigger and better brain. According to many scientists, that seems to be what happened. A 1999 University of California Berkeley press release on the work of physical anthropologist Katharine Milton explains:

> *Human ancestors who roamed the dry and open savannas of Africa about two million years ago routinely began to include meat in their diets to compensate for a serious decline in the quality of plant foods…It was this new meat diet, full of densely packed nutrients, that provided the catalyst for human evolution, particularly the growth of the brain…Without meat, it's unlikely that proto-humans could have secured enough energy and nutrition from the plants available in their African environment at that time to evolve into the active, sociable, intelligent creatures they became.*

Again, proto-humans and early humans didn't just step out onto a savannah and start killing animals. They first had to evolve the physical capabilities we have reviewed. But once they acquired these assets, they had everything they needed to follow a diet that helped to enlarge and improve their brain.

Meat had clear advantages over tubers and other edible plants. For one thing, prey animals were probably highly visible and numerous on the African savannahs where *Homo sapiens* arose. The store was well-stocked. Another advantage: Meat contains nutrients known to stimulate brain growth. They include proteins, amino acids, and vitamins B-12 and B-3 (also known as niacin or nicotinamide). Significantly, the amount of B-12 and B-3 in plants ranges from little to none; a few hundred thousand years ago, the only practical source was meat. A 2017 paper by two British scientists (Adrian Williams and Lisa Hill), tells us:

> *A key brain-trophic element in meat is vitamin B3/ nicotina-mide ... We view human evolution, recent history, and agricultural and demographic transitions in the light of meat and nicotinamide intake ... We will argue that it [competition for calories] was really about an optimal supply of nicotinamide/ nicotinamide adenine dinucleotide (NAD), [which is] important not only for the energy supply, but also for the metabolic and genetic regulation of growth and big brains.*

Other scientists (Jessica Thompson at Yale, for one) emphasize the role of fat, especially as derived from bone marrow, which requires only a heavy rock to access. The authors state,

> *We propose that the regular exploitation of large-animal resources – the 'human predatory pattern' – began with an emphasis on percussion-based scavenging of inside-bone nutrients, independent of the emergence of flaked stone tool use.*

With this idea in mind, eating animal products could have started long before we existed as a separate species able to make tools and weapons. Also, it may be important that soft bone marrow could have been fed to young children, who lack the teeth and jaws to chew raw meat. According to a University of Michigan study that was reported in sciencedaily.com, bone marrow provides vitamin B-12 and is a significant source of the hormone adiponectin, which helps maintain insulin sensitivity, break down fat, and has been linked to decreased risk of cardiovascular disease, diabetes, and obesity-associated cancers.

A question arises: If meat was so important to early *Homo*, why are there millions of healthy vegetarians in the world today? The short answer is that people know a lot more than we used to know about diets. We have a detailed knowledge of what the body needs, and we know where to find it. We know which beans and greens provide the protein we need. We know how to enrich a vegetarian diet with precisely the vitamins and supplements not generally found in plant food. A few hundred thousand years ago, meat and meat products were probably the only practical sources of the specific nutrients that promote brain growth.

### 10. Enjoying the sexual reward

This, of course, was a compelling goal of life itself: to pass on successful DNA to future generations. Sex as a goal is not often discussed in scientific reports on prehistoric human life. But it had to have been a powerful motivator, simply because it's a need that many men and women felt almost every day. Our ancestors hunted, cooperated, shared, ate, and protected females all to reproduce. They didn't have that goal consciously; they were driven by more profound forces than conscious thought. But they did it all well enough to produce successors who would one day control the world. For better or for worse.

## Combating enemies

The violent nature of men may have started with struggles for mating rights among individuals in a group, just as it does with many other species: "my genes, not yours." We can assume it also stemmed from violent confrontations among groups for hunting grounds and sources of clean water. A few hundred thousand years ago, there must have been plenty of animals in Africa, but not necessarily an ample supply of animals that were relatively easy to kill. Animals learn. Over time, a group that might have been easily approached could become wary or aggressive. This would set up confrontations: If a hunting party met another hunting party in its usual territory, a clash may have often followed. The same weapons and techniques used to kill prey animals would have worked against human animals. The winners of such battles were more likely to pass on their DNA than the losers.

It may have been the same with clean drinking water. It is said that a healthy adult human can go for weeks without food, but only three days without water. We can therefore assume that early humans settled near water. But nearby lakes and rivers may have contained giardia, fecal matter, and other harmful stuff. Early humans may have noticed the ill effects of drinking infected water, and this would have prompted them to search for clean spring water. Any group that found such a spring might have been willing to defend it with their lives.

**Summary**
The theory we've just reviewed suggests (but doesn't prove) that no single factor caused the emergence and development of *Homo sapiens*. Instead, it could have been the natural selection that happened as a result of the 10 steps inherent in the quest for meat and clean water. Each step aided survival, and the entire process, working through natural selection, helped to develop the cognitive, cooperative, and aggressive traits that enabled one *Homo* species to become *Homo sapiens*.

# 8

## The Most Brutal Animal?

A bachelor male red deer decides he's ready to take on the alpha male of an established group. They walk parallel to each other and go through other ritualized maneuvers until one attacks the other. The bachelor is younger and faster, but the alpha male is larger and more experienced. Eventually, amid a clack of bare antlers, the alpha forces the bachelor to break off and trot away.

Now here's the critical point: The alpha male does not kill the interloper. He may chase him a short distance, but then returns to his herd. This evolved behavior may rest on the simple fact that there's no point in expending energy and risking injury once his dominance is reaffirmed.

Compare that outcome with the chimpanzee behavior described by Jane Goodall and other primatologists: Over time, the males in one group attack the males in a nearby group, until every neighboring male has been killed, including low-ranking individuals who present no threat. Significantly, females are typically spared, as are some of the infants. The aggression is not total.

Compared with the red deer, there has been an escalation, not so much in the intensity of aggression, but more in sheer viciousness – brutal action that serves little or no useful purpose for the attackers. Gaining additional territory, with its food and females, does provide a substantial gain to the invaders. But the invaders could have achieved this goal without killing all the neighboring males.

We needn't look far for the next step up in the intensity and level of brutality. It comes from us – *Homo sapiens*. In our pre-history and more recently, we have tortured and slaughtered each other by the millions. Why? Why did the Japanese in 1937 murder more than 200,000 Chinese civilians in a frenzy of killing? Why did the Germans in World War II use scarce resources to imprison and murder an estimated 11 million civilians, including women and children – people who were no threat to anyone? Why are the Russians today bent on destroying a country that is no threat to them? In 2023 why did a group of Congolese rebels invade Uganda and kill 40 people in a high school, mostly students? After invading a village, why have soldiers sometimes forced unarmed civilians into a ditch, then killed them all? Why in 2022 were 550 people shot in road rage incidents in the United States? The victims were total strangers to the shooters. Why do we read about vicious shootings in the U.S. almost every day – acts that inflict intense pain and suffering without benefit to the perpetrators? In this bizarre mode, lethal aggression needs no goal.

Summarizing, we can see there are at least three levels of lethal aggression:

- Need-oriented killing, as when a big cat attacks an antelope for food.
- Selective killing, as when male chimpanzees kill all the males in a neighboring group, but spare the females.
- Gratuitous killing, as in incidents of genocide, soldiers killing civilians or prisoners, and road rage. This level of aggression is perpetrated only by humans.

Genocide is more common than is generally realized. In *The Third Chimpanzee*, geologist Jared Diamond lists 37 different genocides occurring from 1492 to 1990, 17 of them since 1950.

How to explain this kind of behavior?

## Extreme actions

One answer may be that several obstacles to human survival goaded our ancestors into extreme attitudes and actions. They include:

### Physical limitations

On one side, 400-pound antelopes with massive necks, sharp-pointed antlers, smashing hooves, and dazzling speed. On the other side, 140-pound adult primates – slow, relatively weak, and totally lacking in the somatic weapons wielded by their prey. What kind of behavior would help to level the killing field?

### The hazards of ancient hunting

As discussed, killing large animals with primitive weapons was dangerous work, even if we ignore our ancestors' physical limitations. Men were injured and killed as they tried to isolate and kill large prey animals. What actions would lessen the danger?

### Exposure to predators

By the time we emerged as a species, we were no longer proficient at living in trees. We had to live on the ground, which was, and still is, a dangerous place. Chimpanzees, bonobos, and orangutans deal with that danger by sleeping in trees at night, secure in nests they build from twigs and leaves. It helps that they are much stronger than humans and very adept at climbing trees. Videos show them scampering up trees with ease, which is one reason they're good at grabbing agile monkeys. Not so members of the *Homo* line. Most *Homo* species had arms that were shorter and weaker than those of other great apes, and our feet, unlike those of chimpanzees, are not suited for grabbing hold of tree limbs. Our ancestors probably stayed on the ground all day and all night. That meant we were vulnerable to hungry predators, like big cats, hyenas, and raiding humans. What attitudes and actions would tend to provide protection?

Our physical limitations, the dangers of hunting, and living entirely on the ground – taken together, these obstacles presented a substantial barrier to our survival as a species. A few hundred thousand years ago, it would have been reasonable to expect *Homo sapiens* to become another extinct species known only to specialized scholars. But our ancestors didn't let that happen.

## Our suite of assets

We had certain assets that enabled us to deal with these problems. With our bigger, super-wired brain, we could make tools and weapons and use them as extensions of our body. We could run for long distances to stay close to prey, sweat to keep our body cool, and throw spears to bring home meat. Further, our apparent ability to talk helped us develop coordinated social and hunting behavior to an extent not available to other animals.

Within a human group, our ancestors were caring and compassionate; this asset strengthened our ability to bond and cooperate. At the same time, males had the hormones and physical assets required to be successful hunter-killers, despite the inherent danger. This ability enabled us to make meat a substantial part of our diet, which, in turn, helped us grow a larger and much more powerful brain. Not incidentally, these same abilities would have allowed our ancestors to gradually eliminate competitive species – as well as many other humans.

Given our physical limitations, we must have looked like easy prey to big cats and hyenas. (In fact, tigers still kill an average of 1,800 humans a year; hippos, about 500.) To escape the jaws of hungry predators, our ancestors probably had another asset: aggressive males, ready and willing to defend the group. We could have used our upright posture and versatile hands and arms to shake branches and throw rocks and sticks – and eventually, spears – to intimidate advancing predators. This would have been behavior the predators did not see in other prey animals and, we can guess, off-putting. We may have looked like prey, but we didn't act like prey. Adding to our repertoire of defenses was our ability to use weapons as extensions of our bodies. There is a video of an adult chimpanzee menacing a younger and smaller chimpanzee. The bullied animal retreats, until it gets an idea: It picks up a branch from the ground and starts waving it at the attacker, who immediately backs off.

But having all those abilities is not the same thing as using them. To feed the group, somebody had to go out on the savannah, find suitable prey, then kill a dangerous animal with pointed sticks – and that took balls, guts, grit, courage – whatever you want to call it – which, in turn, probably required a sex-specific brain structure and hormonal brew.

We will review both in the next chapter.

## Aggression gone wild

In short, our inclination to the extreme form of aggression – gratuitous aggression – may be the result of the natural selection that came from a desperate attempt to overcome the obstacles that threatened to wipe us out as a species. History (and today's news) shows we over-corrected, moving from necessary defensive actions to needless brutality.

One of the first scientists to suggest the aggressive nature of proto-humans was the paleoanthropologist and anatomist, Raymond Dart (1893 – 1988). Dart is justly famous for identifying and describing the Taung Child, which in a 1925 edition of *Nature,* he called *Australopithecus africanus* (southern ape of Africa). (The first word refers to the genus; the second, to the species.) The name stuck. Many fossils of this genus have since been found and identified. In his introductory paper, Dart stated:

> *...it represents a fossil group distinctly advanced beyond living anthropoids in those two dominantly human characters of facial and dental recession on one hand, and improved quality of the brain on the other...At the same time, it is equally evident that it is no true man. It is therefore logically regarded as a man-like ape.*

Its status as "man-like" was crucial. Dart's discovery was at first well-received, but the warm welcome soon cooled. Most paleontologists and anthropologists came to regard the Taung child skull as that of a young gorilla or other type of ape, definitely not "man-like." Other scientists, and then Dart himself, persisted and later found bones and stone flakes in Africa indicating that various species of *Australopithecus* had hunted and butchered antelope and other animals. An article by Briana Pobiner in American Scientist explains:

> *In other fossils at the same site Dart saw evidence of meat-eating, such as baboon skulls bearing signs of fracture and removal of the brain case prior to fossilization, with v-shaped marks on the broken edges and small puncture marks in the skull vault. He concluded the Taung child had belonged to a predatory, cave-dwelling species he described as 'an animal-hunting, flesh-eating, shell-cracking and bone-breaking ape' and 'a practiced and skillful wielder of lethal weapons of the chase.' The concept of the killer ape was born.*

That concept was notably developed by playwright and author Robert Ardrey in three books: *African Genesis*, *The Territorial Imperative*, and *The Hunting Hypothesis*, in which he says: "The hunting hypothesis may be stated like this: Man is man, and not a chimpanzee, because for millions upon millions of evolving years we killed for a living."

Even if we grant the dogmatism of brevity, that statement over-simplifies a complex idea. Yes, killing prey animals and eating their meat were vital to the evolution of *Homo sapiens*. But the plot is much thicker than that. For one thing, recent evidence indicates that other *Homo* species and chimpanzees also killed animals for part of their diet. So "killing for a living" could not have been the only reason we became human. That said, it should be recognized that Ardrey was a pioneer in describing the profound effect that hunting and eating meat had on the development of our species. His work, especially in *African Genesis*, was vigorously attacked by scientists who believed humans are born with a blank slate and that all our propensities are acquired from our culture. It is now widely (though not universally) recognized that the blank slaters were mostly wrong and that Ardrey was mostly right (when modified to allow for natural selection).

## Non-stop wars

An article on sciencealert.com spells out its message in the headline: "Nine Species of Human Once Walked Earth. Now There's Just One. Did We Kill The Rest?" The author is Nick Longrich, of Bath University. Longrich states:

> *By 10,000 years ago, they were all gone. The disappearance of these other species resembles a mass extinction. But there's no obvious environmental catastrophe – volcanic eruptions, climate change, asteroid impact – driving it.*

*Instead, the extinctions' timing suggests they were caused by the spread of a new species ... Homo sapiens.*

Longrich's statement may ignore the principle that correlation does not prove causation. But combined with a wealth of other evidence, it makes an interesting point. Human males have been killing each other in armed combat for many thousands of years. *The New York Times* article mentioned earlier tells us that some humans engaged in armed combat 92 percent of the past several thousand years. (The article comes from the book, *What Everybody Should Know About War*, by Chris Hedges.) News reports since 2003 give us little reason to think that wars have become less frequent.

The first war recorded by an historian occurred nearly 3,000 years ago, between Sumer and Elam, in what was then Mesopotamia. Sumerians, under command of the King of Kish defeated the Elamites and, an historian said, "carried away as spoils the weapons of Elam." This was not the first war or war-like encounter; it was simply the first we know of recorded by an historian.

The Hellenistic Greeks built a highly successful and relatively prosperous civilization; they had no need to expand. They nevertheless invaded a large part of the known world (Alexander the Great was wounded in India, in 326 BC.) The Romans did the same, extending their dominion northward into northern Europe and eastward into Egypt. And always they did it by the sword. Local populations did not want to be taken over by outsiders, so the natives had to be killed or enslaved. The ongoing attitude exemplified the male drive for dominance: "We are more powerful than you, so we will take what you have, and you will obey our orders."

## An extreme example

The Roman siege of Masada in year 73 of the common era is a clear case in point. Various sources cite differing figures; the following account is meant to represent a broad consensus: The invading Romans had conquered Jerusalem and controlled nearly all of Judea. According to an *Encyclopedia Britannica* article, the only substantive resistance came from about 960 Jewish men, women, and children who had fled to Masada, a plateau 1,424 feet (434 meters) high and surrounded by steep cliffs on all sides. It was a seemingly impregnable position, equipped with an ample supply of food and water and men willing to defend it with their lives.

Undeterred, the Romans spent months, and possibly years, assaulting the area with an army of about 10,000 soldiers and a few thousand slaves. First, they built a 2.6-mile wall around the plateau to keep the Jews from escaping. Then they painstakingly built an earthen ramp about 2,000 feet long and possibly 200 feet high (estimates vary). Having made many sieges, the Romans knew what to do next. They built an armored siege tower and hauled it up the ramp. Using a battering ram on the tower, the Romans managed to break through a stone wall the defenders had built. Then came the anticlimax. When they entered Masada, they found nearly everyone had committed some form of suicide, rather than surrender; the only survivors were a few women and children who had hidden in a drain.

The siege required a huge and costly effort by the Romans. For what? Masada was an isolated position with no mineral deposits or significant amounts of farmland. It was not on a trade route; stood in territory that was already captured; and was in no way a threat to the Romans. They could have used the resources devoted to attacking Masada elsewhere, presumably to much greater advantage. But no, they had to subdue this little village explicitly and unequivocally.

For their part, the defenders of Masada probably would have fared better if they had surrendered. Yes, some may have been killed and the rest enslaved. But even that result would have been better than certain death for the entire population. But like the Romans, they (meaning the men) had to have it all their way, or no way at all.

What's going on here, with both sides making irrational life-and-death decisions? One answer is that their behavior was a function of the nature of men: "I must be in control; I must dominate; I must be seen as brave, resolute, decisive – whatever the cost. Otherwise, I will be seen as weak and cowardly – not a man amongst men." As mentioned, this attitude may have come from the natural selection that occurred when our ancestors lived by the spear. As a young man, either you were brave and skillful enough to join the group's hunting party, or you weren't, in which case your status in your group may have shrunk to near zero. And when you did go on a hunt, either you came back with an ample amount of meat and all its rewards, or you didn't.

The flip side of that attitude is to believe that if you can wield power, you are entitled to subjugate others, take what they have, and, if it suits your purpose, kill them.

**Invasions are us**
That attitude ruled, for example, from the fifteenth through the eighteenth centuries, as Portugal, Spain, England, France, and The Dutch Republic sent invaders to build colonies in the Americas, India, and Africa. These countries were followed by Belgium, Germany, and Italy in the nineteenth century as each staked out claims in Africa and elsewhere. The United States joined the violent action by invading Mexican and Native American lands reaching to the Pacific Ocean.

These invasions were characterized by populations with advanced technology conquering populations with little technology.

Throughout known history, such invasions were considered justified and moral. "After all," the thinking seemed to be, "It's not as if the natives are people like us; they're savages, and we are bringing them the blessings of civilization and religious salvation." A typical invasion force included priests bent on procuring "more souls for Christ" (not to mention gold, silver, and land). Natives who resisted were not just blocking "progress;" they were rejecting God Himself, so of course there was nothing wrong with killing the unbelievers, which is how ISIS justifies their mass murders today.

A high level of intelligence does not seem to help. In the nineteen-thirties, on rational analysis, the Japanese should have quit while they were ahead. They had conquered a large part of the South Pacific and even part of China. While the United States clearly objected to this imperialism, it was not ready to go to war over it. But the Japanese leaders had such a powerful urge to dominate, they attacked an obviously larger and stronger country. On similar rational analysis, Germany in the 1930s should have settled for Austria, Czechoslovakia, Sudetenland, and non-violent access to the farmlands of Poland. But again, the urge to dominate by whatever means proved stronger than reason.

The sheer number of invasions makes the point. According to military.wikia.org, military forces have perpetrated nearly 60 invasions just since 1946. And those who invaded were almost entirely men, not women. We can infer that early humans did the same, because *Homo sapiens* males have behaved this way throughout recorded history. We don't just move into a land where others live and then outcompete them; we cooperate to subjugate and kill people in the native population, then take their resources.

## 60 invasions

As discussed, the males of our species often invade and kill. According to military.wikia.org, military forces have perpetrated nearly 60 invasions just since 1946. And those who invaded were almost entirely men, not women. We can infer that early humans did the same, because *Homo sapiens* males have behaved this way throughout recorded history. We don't just move into a land where others live and then outcompete them; we cooperate to subjugate and kill people in the native population, then take their resources.

## Why we go

Combat operations eventually require people down on the ground who take possession of land, and that is typically where most of the fighting and dying takes place. This fact brings up a question: Why would a man allow himself to be put in a position where he could be killed or wounded at any moment?

The answer starts with the brain structure and hormonal balance we've inherited from our hunting ancestors. One effect of this heritage is to enhance the appeal of certain aspects of military action: We like the excitement, the assertion of power, the profound sense of fellowship, the prestige of serving our homeland, the satisfaction of working for something bigger than a paycheck, the chance for heroic action, and perhaps even the opportunity to kill men designated as enemies, which may be the ultimate expression of male-against-male dominance. Of course, the strength of each of these motivators varies with the individual and his circumstances.

There are also things about war we hope to avoid, especially the risk of death or serious injury. But based on the evidence we've reviewed so far, it's clear that our inherited propensity often emerges as a stronger force than whatever deterrents we may face. The triggers for war are easy to pull.

What can we say about those who serve in needless wars, like the invasion of Iraq? The answer in this book is that those who acted with outstanding courage should be honored as heroes; others are to be commended, and all rewarded. It is next to impossible for the average service member to know if a particular military action is needless or not. Even if high-ranking service members had such knowledge, what would happen if most of them refused to obey their commander-in-chief? It could mean that the military is controlling the country, which would signal the end of democratic government. That danger makes it vital to elect leaders who don't thrust their country into needless wars.

## Into the 20<sup>th</sup> century

In June 1944, General George S. Patton, Commander of the Third Army, took the stage to address his men during a war we had to fight. The men were about to attack the European mainland, where defenses had been prepared by Germany's best generals. They would assault those defenses as barely trained civilian soldiers.

Addressing his troops, Patton could have used any of the time-honored motivations: They would fight for their country. They would fight for freedom. They would fight for their leader. But Patton knew his troops – knew they were willing to fight, but also knew they weren't sure they could. Here are excerpts of what he chose to say, as compiled by Charles Province, founder of the George S. Patton Historical Society:

*Battle is the most magnificent competition in which a human being can indulge. It brings out all that is best, and it removes all that is base. Americans pride themselves on being He Men and they are He Men ... We'll win this war, but we'll win it only by fighting and by showing the Germans that we've got more guts than they have ... We're not going to just shoot the sons-of-bitches, we're going to rip out their living Godamned guts and use them to grease the treads of our tanks ... Why, by God, I actually pity those poor sons-of-bitches we're going up against. By God, I do ... We are going to twist his balls and kick the living shit out of him all of the time ... We are going to go through him like crap through a goose ...!*

Patton's speech may be the quintessential statement of male inter-group aggression. It almost had to be. The men he spoke to knew they were relative amateurs facing highly professional German soldiers who had conquered most of Europe. Patton's men were afraid of being killed; his speech helped them believe that they, civilian soldiers, could also be killers.

# 9

## The Biology of Aggression

Identifying the source of our most aggressive and violent impulses depends on where, in a complex process, you decide to look. The neuroscientist Bernhard Bogerts, mentioned earlier, prefers the most basic view – that it starts in the Y chromosome, which contains a gene that prompts testes to form, along with the external and internal male genitalia. Once this happens, Bogerts argues, the path to eventual aggressive or assertive behavior is established.

In human males, aggressive behavior originates in two of the oldest parts of the male brain, the hypothalamus and the amygdala, both of which sit at the bottom of the brain, near the top of the spine. In brain.harvard.edu, Harvard professor Edward Kravitz explains why aggression has been around for millions of years:

> *Aggression is a nearly universal feature of the behavior of social animals. In the wild, it is used for access to food and shelter, for protection from predation and for selection of mates, all of which are essential for survival...In essentially all species of animals, including man, amines [organic bases] like serotonin have been implicated in aggression. Moreover, good evidence supports the notion that a multiplicity of hormones and neurohormones influence this complex behavior.*

## 5 types of hormones

Aggressive behavior is preceded by a complex system that involves the "multiplicity of hormones" that Kravitz mentions. These five are among the most important in terms of promoting aggression: Testosterone. Serotonin. Cortisol. Oxytocin. Dopamine. We'll look at the role of each:

### *Testosterone*

This is the major sex hormone that tends to affect human males. Some studies show that testosterone tends to cause aggression directly; others indicate that it contributes to and aggravates aggression, but doesn't cause it. In *The Trouble With Testosterone*, for example, Robert Sapolsky points out that levels of aggression often plummet in the members of many species when the source of testosterone is removed, a fact suggesting that testosterone causes aggression. But he cautions that testosterone levels predict nothing about who in a group of human males is likely to be the most aggressive. He believes that testosterone isn't causing aggression; it's exaggerating the aggression that's already there.

In another book (*Behave*), Sapolsky refers to the "challenge hypothesis," which contends that rising levels of testosterone increase aggression *only* when a person feels she or he has been challenged. But people are challenged every day: at work, in the gym, playing games, even at home. So from this point of view, testosterone frequently affects how assertive or aggressive people feel.

The issue is further complicated by the complexity of human motivations. The man with the most testosterone in a group may be *inclined* to act aggressively, but may deliberately avoid such action for a variety of reasons: He may be naturally shy and retiring. He may be afraid he'll make a fool of himself. He may not want to offend anyone.

The list goes on. For a better understanding, let's look at two books that advocate divergent views on the effects of testosterone:

- *Testosterone Rex*, by Cordelia Fine, writer and professor of history and the philosophy of science at the University of Melbourne, Australia.
- *T*, by Carole Hooven, scientist and co-director of Harvard's Department of Human Evolutionary Biology.

### Testosterone Rex: Myths of Sex, Science, and Society:

The main thrust of Fine's book seems to be an unrelenting effort to show that women are just as good as men, mainly because they have the same capabilities as men. She does a commendable job of dispelling various myths about the alleged limitations of women compared with men. The book is valuable for that content. Unfortunately, Fine gets carried away by her zeal and makes some unsupportable statements. She admits there are biological differences between men and women, but claims that whatever differences exist have no material effect on behavior or capabilities. She attributes any observable differences to the impact of culture, not biology.

More recent studies indicate that Fine is wrong on at least two major points: Men and women do have many of the same capabilities, but they also have important differences, starting with the brain. This fact accounts for females being generally better than males at some tasks and males being generally better at others. An article in the 2014 *Journal of Clinical & Diagnostic Research* summarizes some of the differences between the brains of men and women.

*Male and female brains show anatomical, functional, and biochemical differences throughout life ... Males outperform females in tests of visual-spatial ability, and mathematical reasoning, whereas females do better in memory and language use. Moreover, females have different mental skills at different phases of the menstrual cycle...Conclusion: Male cognitive functions were comparable to female preovulatory phase cognitive functions. However, females, during postovulatory phase of their cycle, may have advantages in executive tasks (Stroop test) and disadvantages in attentional tasks (VRT), as compared to males.*

One source of different abilities and behavior is a gross difference in the quantity of testosterone. Both human males and females have this hormone, but males typically have 15 to 20 times more circulating testosterone than females, as reported in a 2018 study conducted by the Australian endocrinologist, David Handelsman. More recently – and equally important – Harvard professor Carole Hooven has shown that interactions between the brain and testosterone contribute to behavioral differences between men and women.

In addition, available evidence suggests that Fine errs in stressing culture over biology. In recent decades, studies show that biology often overrides culture in affecting our traits and typical behaviors. In a 1998 article in *Foreign Affairs*, George Mason University professor Francis Fukuyama explains:

*While some gender roles are indeed socially constructed, virtually all reputable evolutionary biologists today think there are profound differences between the sexes that are genetically, rather than culturally rooted, and that these differences extend beyond the body into the realm of the mind.*

Fine's book was reviewed by Sheri Berenbaum, a professor at Pennsylvania State University. While acknowledging that Fine makes valid (and much-needed) points about the capabilities of women, Berenbaum also sees serious flaws. In this statement, she explicitly contradicts a significant part of Fine's thesis:

> *Hormonal differences across men and women are generally considered to contribute to psychological sex differences in two primary ways: During sensitive periods of development, they produce long-lasting changes to brain structure and behavior ("organizational" effects), and later in life, they produce temporary alterations to the brain and behavior as hormones circulate in the body ("activational" effects) ... She does not discuss methodologically rigorous studies of organizational effects, which have shown that exposure to androgens during prenatal development has clear effects on sex-related behavior, including interests and abilities. She thus fails to provide a balanced presentation of the role of hormones in sex-related behavior.*

Berenbaum ends her review with a trenchant observation: "Actions to achieve gender equality do not rest on discarding the evidence for biological influences on sex differences."

### T. The Story of Testosterone, the Hormone that Dominates and Divides Us:

Carole Hooven teaches and co-directs the undergraduate program in the Department of Human Evolutionary Biology at Harvard University. She works as a "hands-on" scientist. At the invitation of Harvard professor Richard Wrangham, Hooven spent 1999 in Uganda studying chimpanzees, a task that included standing under trees and trying to catch chimp urine in a container to measure testosterone levels. She was attracted to her current field by Wrangham and Peterson's ground-breaking book, *Demonic Males.*

The full title of Carole Hooven's book, *T: The Story of Testosterone, the Hormone That Dominates and Divides Us,* signals a robust view of the importance of testosterone. In addition to teaching, Hooven works with people who have experienced the results of testosterone deficiency or excess, both of which typically begin before birth. She was under pressure to soft-pedal study results suggesting that certain aspects of our biology tend to produce behavioral differences between men and women. But, as she says, "the relevant facts about sex differences had little impact on the emotionally charged response" [presumably from people who want women to be the same as men.

Ignoring the noise, Hooven pushed on, perhaps because she could not ignore the hard evidence she had often encountered. Drawing from studies based on sexual selection, behavioral endocrinology, natural anomalies like CAH (a disorder affecting the adrenal glands), and the effects of castration, Hooven has produced a definitive guide to the impact of testosterone on men and women. Among her book's more informative statements:

> *Testosterone changes everything. It changes the way genes are expressed on every chromosome – proteins from thousands of genes are produced in systematically different patterns and quantities in men compared to women. These proteins go on to affect the body and the brain, first in utero, again shortly after birth, and then there's another explosion of changes in puberty.*

*Evidence from experiments in mammals…shows that when females are given high T in utero, their behavior is masculinized, and when males are deprived of it, their behavior is feminized.*

*The powerful evidence from studies on CAH girls [those with a cortisol defect] seems to seal the deal: exposure to high levels of testosterone even before we are born, masculinizes not only our bodies, but also our interests, preferences, and behaviors.*

In *Demonic Males*, authors Richard Wrangham and Dale Peterson provide a cinematic account of one of those behaviors in describing a chimpanzee lethal attack:

*It began as a border patrol. At one point they sat still on a ridge, staring down into Kahama Valley for more than three-quarters of an hour, until they spotted Goliath, apparently hiding only twenty-five meters away. The raiders rushed madly down the slope to their target. While Goliath screamed and the patrol hooted and displayed, he was held and beaten and kicked and lifted and dropped and bitten and jumped on. At first, he tried to protect his head, but soon he gave up and lay stretched out and still. The aggressors showed their excitement in a continuous barrage of hooting and drumming and charging and branch-waving and screaming. They kept up the attack for eighteen minutes, then turned for home, still energized…*

This description of murderous chimpanzees suggests that the attackers were high on something – high in the sense of feeling the effects of a stimulating drug. The evidence reported in this chapter shows that some genes and hormones can contribute to vicious behavior in humans and other animals. Wrangham and Peterson, again in *Demonic Males*, quote a journalist describing what it's like to be in the middle of a British soccer match:

> *They talk about the crack, the buzz, and the fix. One lad, a publican, talks about it as though it were a chemical thing or a hormonal spray or some sort of intoxicating gas.*

An article by David Kelly, published in the *Los Angeles Times,* describes a tournament in Colorado in which combatants dressed in medieval armor batter each other with metal swords and axes. One avid participant was heard to say, "I'm just here for the violence."

Testosterone is only one of more than 50 hormones in the human body. Here is a brief sketch of four more known to affect human behavior:

*Serotonin*

Serotonin is a messenger hormone with many and varied effects, including the regulation of mood, digestion, sleep, and aggressive behavior. Its role is complex and hinges on many factors. In some physical contexts, it acts to reduce aggression; in others, it has the opposite effect. Below-normal levels are often associated with increased aggression.

*Cortisol*

Produced by the adrenal glands, cortisol helps the body deal with stress. It increases sugars (glucose) in the bloodstream, improves the brain's use of glucose and the availability of chemicals that repair tissues. Cortisol also dampens functions that might hamper response in a fight-or-flight situation.

*Oxytocin*

A neuropeptide produced in the brain, oxytocin functions as a hormone or neurotransmitter. Scientists (Carsten De Dreu, Lindred Greer, and others in PNAS) have found that it "motivates in-group favoritism and, to a lesser extent, out-group derogation. These findings call into question the view of oxytocin as an indiscriminate 'love drug' or 'cuddle chemical' and suggest that oxytocin has a role in the emergence of intergroup conflict and violence."

Regarding the divergent roles of oxytocin, the concept of "in-groups" and "out-groups" may apply. These terms refer to a well-established tendency among many social animals to define themselves as members of a specific group and to define all others as outsiders, which can lead to hostile ethnocentrism.

The concept was developed by several scholars in the early 1900s, most notably by the social and economic Darwinist, William G. Sumner. Actions that strengthen the in-group, like altruism, have a mirror image in out-group aggression, which can also strengthen the in-group by fending off a take-over or reducing competition for resources. In addition, individuals who implement the double role promote their self-interest: They enhance the perpetuation of their own genes, or genes they share with other members of the in-group.

As Johan van der Dennen and others have pointed out, the definition of another group as "not us" can provide societal and personal permission to kill a member of one's own species. With the "not us" attitude in control, an individual is free to behave according to the biochemicals that have an influence on aggression. Richard Wrangham's most recent book, *The Goodness Paradox*, elucidates the significant point that individuals can be compassionate with members of their ingroup and brutal with members of outgroups. Which is why conflict-driven propaganda often exaggerates the differences between warring nations.

*Dopamine*

Males of many species need to feel powerful and energized to maximize their mating opportunities. To help achieve that goal, anger and aggression can produce a flood of biochemicals that may be experienced as feeling good. Psychiatrist Anthony Storr describes this process in one of his books, *Human Aggression:*

> *In other words, a circular reaction is set up in such a way that the brain, which initiates the emotional response, is itself stimulated by the reaction ... The important point is that the body contains a coordinated physico-chemical system which subserves the emotions and actions which we call aggressive, and that this system is easily brought into action both by the stimulus of threat, and also by frustration.*

Storr's analysis seems to be supported by a Physorg.com article that summarizes work at Vanderbilt University by Craig Kennedy and Maria Couppis. Their research shows for the first time that the brain can process aggression as a reward – much like sex, food, and drugs – offering insights into our propensity to fight and our fascination with violent sports like football and boxing. "It is well known that dopamine is produced in response to rewarding stimuli such as food, sex, and drugs of abuse," said Maria Couppis, who conducted the study as her doctoral thesis at Vanderbilt. "What we have now found is that it also serves as positive reinforcement for aggression."

Two other hormones, epinephrine and norepinephrine (also called adrenalines) help equip the body for fight or flight. They have been linked to aggression, but not in a causal role. A downhill skier, for example, might have a high level of epinephrine without the slightest intention of hurting anyone.

## About mice

Clearly, there is plenty of evidence suggesting a biological basis for aggression. But most studies on this topic have been done with mice. It's fair to ask if studying the behavior of laboratory rodents tells us much about the behavior of humans. The National Human Genome Research Institute website provides an answer:

> *Overall, mice and humans share virtually the same set of genes. Almost every gene found in one species so far has been found in a closely related form in the other. Of the approximately 4,000 genes that have been studied, less than 10 are found in one species but not in the other.*

## A passionate debate

All of which brings us to a discussion of which is more important in shaping behavior: The prevailing culture? Or inherited genes and their effect on hormonal balance? There are at least three major views on this general subject:

## Culturism. Sociobiology. Evolutionary psychology.

Culturists hold that human behavior derives far more from culture than from genetic inheritance. Fine's views, as expressed in *Testosterone Rex*, provide a good example of this kind of thinking. From a perch as the most popular way to explain human behaviors, culturism has recently declined in scholarly acceptance.

Many researchers now maintain that genes are more important than culture in influencing behavior. The most prominent figure in that camp is the former Harvard biology professor E. O. Wilson, author of *Sociobiology: The New Synthesis*. This book is generally regarded as a monumental work that created a new field of scientific study. Wilson defines sociobiology as "the systematic study of the biological basis of all social behavior." The theory explains social behavior as a product of natural selection and emphasizes the role of genes in stimulating behavior that maximizes their chances of reproduction. In 1989, officers and fellows of the International Animal Behavior Society rated *Sociobiology* the most important book on animal behavior of all time. Some of Wilson's comments in the book:

> *Genetically based variation in individual personality and intelligence has been conclusively demonstrated, although statistical racial differences, if any, remain unproven... given the overwhelming evidence at hand, the hereditary framework of human nature seems permanently secure.*

Wilson does not ignore his critics from the "blank slate" school of thought, which maintains that biology can be ignored, or at least, minimized, in explaining individual differences in human nature and behavior. In response, Wilson says:

> *No serious scholar would think that human behavior is controlled the way animal instinct is, without the intervention of culture ... To suggest that I held such views, and it was suggested frequently, was to erect a straw man – to fabricate false testimony for rhetorical purposes.*

A third way of explaining social behavior is by the theory of Evolutionary Psychology. Like Sociobiology, it attributes social behaviors to natural selection helping an organism pass on its genes to the next generation. It differs from other explanations of behavior in that it assumes the existence of inherited adaptive mechanisms that reside in the brain of living organisms and influence their behavioral traits. Researchers working in this field focus on the problems they believe our hunter-gatherer ancestors faced, and on the kinds of behavior they developed to solve those problems. Examples of mechanisms include ways of getting food, language development, incest avoidance, mating preferences, and developing alliances. Like sociobiology, the theory has both staunch adherents and passionate critics.

## The effects of gender and aging

Experimental results indicate that men and women, while similar in many ways, are also significantly different. They are different not just physically, but also different in some of their typical behaviors. It's helpful to differentiate between what each gender is *inclined* to do – what each has a propensity to do – and what each *can* do. With sufficient training, some women can crawl through mud, evade enemy fire, throw a grenade, and charge a machinegun nest. But history tells us that engaging in infantry combat comes more readily to most men than to most women. Reversing the example, it's likely that most men respond to a baby crying in a public place by wishing someone would tend to it; we can guess that most women would want to get up, find the baby, and comfort the baby personally or find its mother.

That said, we have to be careful to avoid letting a knowledge of propensities in some situations spill over into areas where it doesn't belong. For example, there is no biological trait that keeps a woman from managing a business – or a country – every bit as well as a man. In fact, as we'll see later in this book, there are reasons to believe many women can manage organizations better than many men.

The biological difference between men and women is at its height in young adults. After menopause, the behavioral differences between men and women may shrink to a lower level. This fact is important, because it is usually not until a person reaches the age of about 35 that he or she has the experience and maturity to manage a large organization. Societies that want more women in positions of power need to establish an array of programs to help women while they're still young so they can prepare for executive positions (see the last chapter of this book, where these programs are outlined).

# 10

## Why Are We Alone?

In previous chapters, we saw that our ancestors were aggressive hunters who cooperated to kill large animals with primitive spears. Their Quest for Meat helped them become human. But that does not explain why the resulting primate replaced all other primates.

The answer may start with the fact that humans were (and still are) the most aggressive and most cooperative of all primates. No other primate has murdered millions of its own species; no other primate has developed anything like the human level of social and technological cooperation.

To understand the full impact of extreme aggression coupled with extreme cooperation, we need to realize that aggression by itself would not have been effective against fast, nimble, and powerful animals; it was more likely to have resulted in failed hunts and injuries to the hunters.

Cooperation by itself would also have been ineffective. Yes, a group of early humans could coordinate spear-throwing from a distance, but that would only wound a large animal like a zebra or eland antelope. To make a kill, the hunters had to get close enough to an animal to sink a spear into a vital organ or cause massive bleeding. Our ancestors probably made kills by surrounding and confusing their prey. Using their big brain, alpha male humans may have shouted orders to enhance cooperation. As described earlier, this close work required a high level of aggression, similar to what a linebacker needs to punch through an offensive line in American football.

## The power of synergy

But just having these traits is still not enough to explain our exclusive dominance. Several proto-humans were also bipedal hunters and had larger brains than prior *Homo* species. Why aren't we sharing control of the world with one or more of these species? Neanderthals, in fact, had an average brain size larger than that of some humans and may have been able to produce language sounds. It can be argued that our unique ability to survive and dominate came from a combination of a big brain, cooperation, and assertion/aggression:

Our brain helped us decide what to do.
Cooperation made it physically possible.
Assertion and aggression made it happen.

This, it can be argued, is how an otherwise ordinary primate prevailed over and, perhaps, dispatched all other *Homo* primates. The combination was synergistic, surpassing what might have been expected simply by totaling the separate effect of each component. It produced a super species, willing and able to control the world.

The ability of *Homo sapiens* to dominate all other animals has been continuously fostered by "serial cooperation," which allows us to preserve acquired knowledge by writing it down and accessing it over time. Cooperating in this way could have been driven by a mutation enabling symbolic thought. If so, natural selection would have conserved the relevant strands of DNA. Serial cooperation has helped thousands of generations of humans make continual improvements in weapons, product production, distribution, and lifestyle.

## Cooperative killing

As author Robert Ardrey observed many years ago, human ancestors often engaged in killing prey animals and sometimes other species of

*Homo* and other humans. We have reviewed a large body of evidence indicating that the aggressive and violent behavior of *Homo sapiens* males was passed on by natural selection, which also conserved our tendency to be compassionate with ingroup people and hostile to outgroup people.

Many male humans have retained the biological basis for cooperative killing, because little has changed in human evolution to eliminate the advantages of cooperation and aggression. These traits were adaptive while humans developed as a species; they have remained adaptive throughout human history and are still adaptive today.

## From hunting to war

In *How War Began*, Texas A&M professor Keith Otterbein asserts a connection between killing other animals and killing our own kind. He states: "The approach I have taken in this book shows, through an extended narrative, how hunting led to warfare." He cites four parallels:

- Same weapons.
- Same search and kill methods.
- Same level of coordination.
- Same need to range far, which leads to encounters with other hunters, who don't want strangers in their territory.

To support his theory, Otterbein presents a statistical analysis of data on subsistence tribes from several researchers. He states:

> *We can conclude that bands that depend upon hunting for subsistence have more warfare than those with little hunting and also that bands that have great dependence upon*

*gathering have less warfare than those that get only a small part of their subsistence from gathering.*

Other scientists dispense with the connection to hunting. Steven Leblanc, for example, is the former Director of Collections at Harvard's Peabody Museum. In *Constant Battles: Why We Fight,* he makes the case that humans have always engaged in what amounts to war, whether they were active hunters or not. LeBlanc argues that wars tend to happen when a local population grows so large, it outstrips available resources, and further, that an imbalance between population size and resource supply happens frequently, thus fomenting frequent wars. History tells us that a government's desire for resources is one of the motivations for invasions and war. But such practical objectives do not explain why ordinary men are willing to risk their lives in brutal combat that inevitably includes massive death and destruction. All too often, we seem to prefer using violence even when peaceful means of achieving national goals are available, like those discussed in Robert Gates's Symphony of Powers.

**The games we play**

Some people say that when there is no war, we revert to our basic, peaceful selves. Is that true? The implications of the most popular American sport provide a clue:

Imagine a high school football team training for an upcoming game. You don't need to see it; you can hear it: young males shouting in unison, again and again, synchronizing their drive to fight as a team. The sound is visceral: grunts grown loud. Before the game, ancient rituals reappear: Young people rally to stoke excitement. Females cheer to stimulate the males. Honor is paid to the combatants. Minutes before the violence begins, the leader exhorts his players to fight with all their power – they must triumph over the opposing team.

Charged with a flood of feel-good chemicals, the young males take to the field. A roar of approval fills the air. The home team's name is chanted. Drums pound. Music pulsates. Females dance. A fervid celebration of testosterone sets the stage for intense combat – and we love it.

We also love movies and television shows laden with violence. The modal movie poster shows a person with a gun. It could be argued that this preference is just another example of learned behavior. But this claim misses a vital point: Team owners and entertainment producers are not trying to influence public attitudes one way or another. They just want to make money, so they offer what people will pay for.

Evidence of aggression is pervasive in our culture. Cars and trucks are designed to have an "aggressive" look. A relatively small business enterprise with an impact larger than expected is said to "punch above its weight." Sales prospects are referred to as "targets." Some large companies have carried aggressive behavior to unethical extremes: The severe recession in 2007 and 2008 was caused mainly by giant banks and other mortgage originators accepting (and promoting) mortgages they knew to be of low quality. They were able to do that, because they also knew they could bundle the garbage and sell it to customers secretly seen as prey.

## Wallowing in aggression

Today, spear-throwing is largely confined to track and field events. And yet, most human males seem to carry the same biochemical brew, which stimulates violent sports like American football, rugby, and ice hockey. This kind of behavior doesn't threaten society, provided it stays on a game field. But it doesn't.

Forty percent of Americans own at least one gun. The Brookings Institution estimates that Americans own 450 million guns, compared with a population of 330 million, giving us more guns than people. The excessive number of guns seems to be a strong contributor to deaths by gun: Per capita rates of gun deaths in U.S. states with a high level of gun ownership are two to three times higher than in states with a low level. In 2022 assault rifles were legal in 40 states. According to *USA Today*, road rage shootings in the U.S. roughly doubled from 2018 to 2022. Also in 2022, a heart-rending statistic: More children died from gunfire than from any other cause.

The problem is larger than the number of guns – we have a gun culture: "Got a dispute? Get your gun." Open carry is legal in more than 30 states, which means guns on our streets, guns in our stores, guns on college campuses. "Stand-your-ground" laws, which may allow a gun carrier to shoot a person he *thinks* is threatening, are on the books in more than 30 states. Aggression rules.

## A ubiquitous trait

We have already covered the biological and historical evidence supporting the view that human males inherit a propensity for violence and war. This discussion centers on ubiquity: If the vast majority of individuals in a species shows a given trait in a broad variety of situations, it suggests the trait is probably inherited, even if it is influenced by culture. Here are nine examples of the ways in which men's propensity for cooperative killing is ubiquitous:

### 1. Political geography

If human males tend to inherit a propensity for lethal group aggression, we would expect to find evidence of this trait in many places throughout the world in any given year. According to Geneva-academy.ch, there were 110 armed conflicts between dozens of countries in 2022.

## 2. Long-term history

As discussed, war has been part of human history for many thousands of years. In *Sex and War*, Malcolm Potts and Thomas Hayden report projectile points in skeletons 20,000 to 35,000 years old. The authors also report other evidence of combat extending back thousands of years in various parts of the world. As mentioned earlier, in *What Everybody Should Know About War*, Chris Hedges reports that at least 108 million people were killed in wars in just the 20th century. Hedges also reports that over the past 3,400 years, humans have been entirely at peace for only eight percent of recorded history.

## 3. Natural selection

A plentiful supply of meat was essential to the cognitive development of *Homo sapiens*. Natural selection, which includes sexual selection, would have favored aggressive hunters, who probably brought home more meat and had more offspring than timid hunters. Over many thousands of years, this process would have tended to produce a species dominated by aggressive males who work in teams. Significantly, this kind of selection has continued even to the present day: Aggressive men tend to reach positions of power and wealth. It is hard to find any event in human evolution that might have changed that pattern.

## 4. Political structure

Greece may have been a democracy and Rome a republic, but that did not prevent their citizens from invading dozens of countries and then killing thousands of people. That kings and authoritarians have often gone to war needs no documentation here. In modern times, even leading democracies – the United States, the United Kingdom, and their allies – have invaded other countries, sometimes without having been directly attacked.

## 5. Population size

Highly populated countries like Russia, England, Japan, and the United States have waged many wars, and so have small countries in the Balkans and Central and South America. In Asia, too, large and small nations have waged war in ancient and modern times. As documented earlier by Keith Otterbein, Steven LeBlanc, and others, many small, indigenous tribes also engage in lethal combat.

## 6. Type of economy

Some warring countries in the past have had centralized economies controlled from the top by monarchies and dictators. But the arrival of democratic capitalism during the Industrial Revolution did little to slow the pace of war. Shooting wars are notoriously profitable for some industries and some people. Collectively, capitalist countries in Europe, the Americas, and Asia have waged many wars, and so have some communist and socialist countries.

## 7. Standard of living

History shows that a group's standard of living is irrelevant to its willingness to wage war. First World nations have fought dozens of wars; so have Third World nations. First World countries want to protect their wealth (often called "national interests"); Third World countries want to gain wealth, or see military action as a way of strengthening the government's hand. Both employ elaborate propaganda and legal measures to mobilize their citizens. A 2020 example is North Korea, which has sacrificed its citizens' standard of living for arms as an assertion of power.

## 8. Level of technology

Whenever an innovative technology appears that can be weaponized, it is weaponized. Examples include: stone points on spears, spear-throwing atlatls, bows and arrows, rifles, and cannons. In modern times, we have machine guns, radar, jet fighters, rockets, drones that make target decisions, and, of course, nuclear weapons, which can now be delivered by nearly unstoppable hypersonic rockets bolting through space. Unfortunately, little respite is found in societies with primitive technology. One example: In *How War Began,* Keith Otterbein shows that even the most primitive and isolated people in many tribes practice a form of war that resembles infantry squad tactics, often with high casualty rates.

## 9. Religion

Judging by the historical record, a country's religion has done little to keep it from killing huge numbers of people. During the Inquisition, Spanish Christians killed thousands of people they defined as heretics. Later, they killed hundreds of thousands (some say millions) of Aztecs and Incas by combat and imported diseases. Christians in the U.S. killed tens of thousands of Native Americans, directly and indirectly. Ottoman (Turkish) Muslims killed close to a million Armenians. German Christians killed six million Jews and five million others, in addition to the millions killed in war. Japanese Shintoists killed over a million people before and during World War II. Khmer Rouge Buddhists killed about one million of their own citizens. Christian Serbs killed thousands of Muslim Bosnians. Hutu Christians killed hundreds of thousands of Tutsis. Muslims in the Mideast have killed tens of thousands. Most of these examples do not involve trained armies confronting each other on a battlefield; they were simply murders. And the killing goes on.

In *The Most Dangerous Animal*, author David Smith talks about "democides," which are government-sponsored killings apart from warfare. He lists 19 that have occurred since 1904 with dates and details, followed by mention of more than 20 ethnic groups that were also targeted for elimination.

Combat and killing occurring across all these metrics suggests that men's ancient propensity for cooperative killing is part of our nature. Look at any place, at almost any time since *Homo sapiens* evolved, with any kind of political or economic system, with any kind of culture or religion, and you are likely to find cooperative killing – of prey and fellow humans.

# 11

## Skeptics Considered

Disagreements with the central points of this book tend to focus on two points: the nature of chimpanzees and the nature of men, especially concerning our affinity for war. We'll start with chimpanzees.

**The nature of chimpanzees**

As mentioned earlier, only chimpanzees and humans have patriarchal, male-bonded communities characterized by male-driven, lethal raiding of neighboring groups.

Some say the number of chimpanzees killed by other chimpanzees is relatively low. But this arithmetical approach misses the main point. It's not how many animals are killed or wounded over a decade; it's *how* they are killed: Members of a species that shares a high percentage of our mitochondrial DNA go on patrols and raids with the aim of killing other members of their own species and seizing their territory, food, and females – a practice we can read about in the daily news, except that the perpetrators are not chimps. They are us.

According to primate ecologist John Mitani, chimpanzees in Ngogo go on patrol every 10 to 14 days. These are not happy hikers. They steal through the jungle, with the obvious intent of surprising one or more animals from a neighbouring group. When they encounter a "foreigner" in or near their territory, they attack it viciously, sometimes tearing off skin and testicles. The attacks often continue after the victim is clearly dead, indicating the attackers gain pleasure from the attack itself.

Note: It's important to distinguish between patrols, which are relatively frequent and usually skirt the edges of a group's territory, and raids, which are infrequent and require deliberately invading another group's territory.

Chimpanzee patrols and raids are strikingly similar to human infantry patrols. They have the same tactical goal (find the enemy in contested territory and kill him), the same methods (advance quietly in single file), the same strategic goals (assess the enemy's strength and prevent enemy encroachment). And like human males, chimpanzees commit what humans regard as atrocities, as established by Wrangham and Peterson in *Demonic Males*.

Our knowledge of wild chimpanzee behavior comes from researchers who have spent many years in African jungles observing these animals. Other scientists have questioned the validity of their findings. They say that the reported behavior – especially the killing behavior – could have been driven by human intervention in the form of habitat loss and artificial feeding. To address this concern, 30 researchers in 2014 compiled data sets that covered 50 years of research in 18 chimpanzee communities. Ian Gilby, codirector of a Jane Goodall research unit and an Arizona State University anthropologist, summarized the result in a 2014 ASU press release: "This study debunks the idea that lethal aggression among wild chimpanzees is an aberrant behavior caused by human disturbances, like artificial feeding or habitat loss." David Morgan, a research fellow and study participant, echoed Gilby's conclusion in a ScienceDaily post: "We found human impact did not predict the rate of killing among communities." Decades of field research confirm the point: Hunting monkeys for food and sometimes killing neighboring chimpanzees should be seen as natural behavior for chimpanzees.

Culturists who want to believe humans are naturally peaceful sometimes point to bonobo behavior, which is known to be less violent and, some claim, not violent at all. Since bonobos are in the same genus as chimpanzees, why aren't they just as legitimate a model for human behavior as chimpanzees? The answer involves at least two points: First, University of Southern California professor Craig Stanford, who has spent many years studying chimpanzees and bonobos, tells us that bonobos are not as peaceable as is generally thought. In a 1998 *Current Anthropology* article, he states: "While there are marked differences in social behavior between these two species, I argue that they are more similar behaviorally than most accounts have suggested." Supporting that view, a newscientist.com article reports a quote by the Max-Plank behavioral ecologist Gottfried Hohmann, who has studied bonobos in the wild: "Bonobos are merciless. They catch it [a monkey or small antelope] and start eating it. They don't bother to kill it."

The second point about using bonobos as a model for human behavior highlights a fundamental difference: Bonobo society is controlled by coalitions of females; chimpanzee society is controlled by coalitions of males who invade and kill their neighbors. Human society is also controlled by males, many of whom also invade and kill. Taking these points into account, the facts require that we compare humans with chimpanzees, not bonobos.

**The nature of men**

As discussed, some people believe our propensity for cooperative killing is culturally driven, not inherited. They tend to be followers of John Locke, the seventeenth-century philosopher who advocated the idea that the human mind at birth is a *tabula rasa,* or "blank slate." Locke believed that human behavior is driven by data and the ability to acquire it from the culture in which a person develops. As mentioned earlier, this point of view has been weakened by many discoveries in genetics, epigenetics, biochemicals, and studies of identical twins.

As a group, culturists tend to define war in terms of major battles between opposing political entities. They seem to believe this definition allows them to present war as relatively infrequent and relatively recent. But their definition falls short of being a useful way to measure the frequency of armed conflict. Many battles, including those with a high percentage of casualties, occur between groups that do not represent an official state, as we've seen in civil wars and other kinds of conflicts. It's also true that many small-scale actions occur even in large-scale wars; attacking units often use squad-size tactics to look for weak points in their enemy's defenses.

In addition, the culturists are so determined to deny a genetic basis for intergroup aggression, they ignore a basic point: The source of culture must be biology. The learned behavior that culturists refer to does not come to us *deux ex machina*. Short of resorting to religious doctrine, culture appears to originate in the operation of biology.

Another anthropologist in the culturist camp tells us humans often create great mayhem, but also have a remarkable capacity for working out conflicts without resorting to violence. Of course. No one disputes that. But this statement does not address the biological inheritance that promotes lethal group aggression. This anthropologist says he was able to locate "over seventy" non-warring cultures, a point which he offers as proof that, in his words, not all societies make war. He does not tell us the number of people in each tribe, but it's clear that most of the tribes in his list are small, isolated, and primitive, indicating that all these people taken together comprise a tiny percentage of the world's population. Yet he appears to use this atypical sample as a basis for evaluating the rest of humanity.

Before his death, anthropologist Keith Otterbein was recognized as a leading authority on the anthropological aspects of war. Contrast the "seventy non-warring cultures" mentioned above with the following excerpt from Otterbein's 2004 book, *How War Began:*

> *Of all the varieties of human societies, the least likely to engage in warfare are the hunting and gathering bands. This assertion, of course, does not mean that they do not have war. Other types of societies, all socially and politically more complex than bands, almost invariably have warfare. It is rare to find a tribe, chiefdom, or state that has not engaged in war within its recent history. In a cross-cultural study of warfare I found only two such societies in my randomly chosen sample of fifty societies.*

Both of the tribes Otterbein cites are small and isolated. The larger point here is that it doesn't matter how many little tribes have, or have not, engaged in combat. What matters is that men throughout the world have been killing each other in massive numbers for thousands of years and are still doing it today.

Another anthropologist tells us that in most cases the decision to wage war involves the pursuit of practical self-interest by those who actually make the decision: Leaders often favor war, because war favors leaders. However valid this position may sound as an explanation for war, it is logically incomplete: That leaders and their cohorts may seek gain for themselves does not set aside the possibility that they are also driven by an inherited propensity for aggression. Also, we must ask what drives the people in the leaders' domain to risk their lives and, all too often, lose their lives? There is no doubt that cultural factors can be powerful motivators. But much of that power exists, because it rides on the back of inherited drives. Elsewhere, the same anthropologist asserts that war (as he defines it) is a relatively recent invention. He states that many archaeological indicators of war are absent until the development of a more sedentary existence and increasing socio-political complexity.

Most of this view is contested by Keith Otterbein in *How War Began,* where he tells us: "Early hunters working as a group in pursuit of game sometimes engaged in attacks upon members of competing groups of hunters."

As discussed earlier, Harvard archaeologist Steven Le-Blanc also emphasizes the pervasiveness of combat in ancient times. In his book, *Constant Battles: The Myth of the Peaceful, Noble Savage,* he states:

> *The common notion of humankind's blissful past, populated with noble savages living in a pristine and peaceful world, is held by those who do not understand our past...The warfare and ecological destruction we find today fit into patterns of human behavior that have gone on for millions of years...I have only recently come to realize that wherever I had dug, regardless of time period or place, I have discovered evidence of warfare...We are much better off understanding the reality concerning warfare and human ecology, and getting this right is very relevant to understanding how humans became humans and how we function as humans.*

That reality is still being questioned by some scholars, as in the example below: In the August 2020 issue of *Scientific American,* Duke professor Brian Hare, and Vanessa Woods, scientist and director of the Duke Puppy Kindergarten, wrote an article entitled, "Survival of the Friendliest." One excerpt:

> *Compared with other human species, it turns out we were the friendliest. What allowed us to thrive was a kind of cognitive superpower: a particular kind of affability called*

*cooperative communication. We are experts at working to-gether with other people, even strangers ... This friendliness evolved through self-domestication. Domestication is a pro-cess that involves intense selection for friendliness."*

Taking a similar view, some prominent scholars point out that humans were more aggressive and violent in the past than in recent times. That may be true, even though 19 million U.S. citizens are licensed to carry a concealed gun, some of whom probably attended the seditious attack on the U.S. Capitol on January 6, 2021. Still, it can be argued that most men do not walk around with weapons, ever ready to engage in mayhem. We no longer hear of men fighting ritualized duels, as Alexander Hamilton did with Aaron Burr. There can be plenty of aggression on the job, but when a disagreement occurs, we generally don't start bashing each other over the head, and that may represent a change from how things used to be.

But is "friendliness" the best word to describe a situation where people work with one another, with little regard for how they feel about each other? Or would "shared-goal cooperation" be a more accurate term? Early on, hunters must have realized they had to work as a team to bring down large animals with crude weapons. Later, it must have been clear that, to build a bridge or construct an aqueduct, cooperation was *sine qua non* – you can't build a bridge with an enemy. So, yes, it appears that some level of self-domestication and friendliness must have occurred in the history of our species, especially in some European cultures. But it's hard to see how it could have been more formative than the natural selection that flowed from 10,000 generations of hunting and killing large animals.

## Are we the friendliest?

Taking an overview, some of the points we've covered cast doubt on the friendliness theory:

*Ingroup vs. Outgroup*

In his book, *The Goodness Paradox*, Harvard professor Richard Wrangham points out that people are often compassionate and supportive of those they regard as "one of us," a point that supports the friendliness theory. He also observes that those same people can be (and often are) vicious and brutal with those perceived as belonging to an outgroup.

*Perpetual war*

Men have waged war almost continuously throughout recorded history. More than 100 million people have been killed in wars in the twentieth century alone. As we saw earlier, the danger from advanced military weapons has increased, not subsided.

*An inherited inclination*

From the evidence we've reviewed, it's reasonable to infer that most men inherit a propensity for combat and war. This observation comes with the caution that a propensity is an inclination, not a command or instinct.

*Natural selection*

Our propensity for combat probably stems from the natural selection that occurred during nearly 300,000 years of hunting and killing fast and nimble animals with crude spears. Hunters who returned with meat were more likely to reproduce than were empty-handed hunters. Hunters who cooperated, friendly or not, were probably more successful than those who did not cooperate.

*The importance of meat*

The qualities needed for successful hunting had to be more important in human evolution than friendliness, for a basic reason: Meat was biologically essential to the development of the human brain. There is no known evidence indicating that friendliness has this capability.

*A gender-dependent difference*

The friendliness theory fails to account for the behavioral differences between males and females. Throughout history, it has been men who start wars and fight wars, not women. As cited earlier, research shows that when women are aggressive, they tend to show it in non-physical terms.

In short, a review of the available evidence indicates cooperation was important in human evolution, but not friendliness as the term is generally understood.

## Diversity not a cause

Some researchers appear so intent on denying men's biological inclination for combat and violence that they dispute the implications of their own research. Example: In a 2020 *Scientific Reports* article, the authors start by acknowledging the violent history of our species:

> *The existence of a non-violent past is now considered disproved, as aggressive behaviour has generally been demonstrated in both prehistoric and modern hunter-gatherer communities.*

The researchers' own fossil discoveries in the Spanish Pyrenees reveal the massacre of a group of farmers around 5,300 years before the common era. But then the authors say:

*On closer inspection, it seems that it is less our nature than our cultural diversity that impedes universal peaceful coexistence. A sustainable future will therefore only be achieved through mutual respect, tolerance and openness to multiethnic societies as well as the elimination of barriers between cultures and religions by on-going dialogue.*

With this statement, the authors imply that male aggression is not biologically prompted and that the route to peace lies in reducing the problems that come with cultural diversity. It does seem that differences in religion, race, and other factors can aggravate strife. But if cultural diversity is the primary cause of war, how do we explain the large number of civil wars, where groups with similar cultures and interests fight each other? Based on a criterion of 1,000 casualties per year, there were at least 200 civil wars from 1816 to 1997, roughly half of which occurred from 1944 to 1997. A telling example is the American Civil War, which, according to history.com, caused the death of 620,000 American soldiers, roughly 215,000 more than died in World War II. The combatants had many similarities:

- Same race (Caucasian)
- Same language (English)
- Same general religion (Christianity)
- Same legal system (British)
- Same cultural inheritance (British)
- Same economic system (capitalism)

Nevertheless, the opposing forces slaughtered each other by the hundreds of thousands. Multiple similarities failed to dampen the male appetite for combat.

The war in Afghanistan offers a more recent example. According to a 2020 article in *The New York Times*, about 50,000 Afghan policemen have been killed by resident Taliban over the past five years, while the Taliban fighters themselves have lost at least an equal number. The article quotes an Afghan military officer as saying, almost wistfully, "Who is it on the other side?" I wish it were people from a different country that we were fighting — they are not even from a different district." The combatants are nearly all native Afghans, speak similar languages, and share the same culture. The only difference between them is that one side professes a different version of their shared religion. Clearly, cultural diversity may contribute to war, but it cannot be the main cause.

## The fuel of war

Observers and scholars like Carl von Clausewitz and the English historian, John Keegan, make valid points about the causes of war. The usual suspects – greed, sociopathic ambition, fear of foreign action, competition for resources – often act as catalysts; but in the end, it appears that war is not to be understood by abstract or statistical analysis. Instead, policymakers need to realize that the engine of war is fueled by the biological nature of men. With rare exceptions, it is men who make war and men who fight wars. Not women. We have covered the reasons, which flow mainly from the natural selection that accompanied the male history of killing large animals with primitive weapons. Judging by fossil finds and other evidence, we went from cooperatively killing prey to cooperatively killing other species of *Homo*, as well as other humans. Taking all these factors into account, the logical inference is that men operating without women cannot be trusted to avoid war, even when war is not necessary for national security. To offset the lingering male propensity for war, the world needs women to share power with men on an equal basis.

**Lift 300 pounds? Nobody cares.**

We need to keep our eye on a crucial fact: For thousands of years, men have dominated and abused women, only because human males are usually bigger, stronger, and have far more circulating testosterone. But those qualities mean little or nothing in a civilized society, where the world of work does not ask how much you can bench-press.

On average, women have different styles, propensities, and cognitive abilities from those of men, but their inherent capabilities for productive work and achievement in civilized societies are the same. The more a society's culture reflects this fact, the better the society will function. The last chapter of this book outlines what must be done to give women a fair shot at assuming political power equal to that of men.


**Summary**

*Homo sapiens* males have combined extreme cooperation and extreme aggression to create a super species willing and able to dominate the world, a fact that enabled the elimination of other *Homo* species. The Quest for Meat played a central role in human evolution by integrating hunting, sharing, and cooking into a brain-building process implemented with the help of testosterone and other androgens. Studies show that hunting, especially of large animals, increased our ancestor's affinity for armed combat. Ancient weapons and spear points in fossils indicate that humans have engaged in armed combat for at least tens of thousands of years.

Armed combat has been pervasive across a broad array of metrics, indicating a biological, as well as cultural, basis for war. Men have never been satisfied with any level of kill-power, always pushing for more, blind to the fact that obsessive weapons development is often a zero-sum game. The use of artificial intelligence makes war more likely by expediting the operations needed to launch military action. To move the world toward peace, women need to wield at least as much political power as men.

# 12

## Male Problem, Female Solution

As discussed, a wealth of evidence shows that it is typically men who decide to wage war – aggressive men driven at least in part by inherited genes, testosterone, and deep-seated cultural attitudes; men who feel they must be seen as strong, resolute, victorious; men advised by similar men with similar attitudes; men whose predecessors have an almost unbroken history of waging war.

Considering their bias toward war, should men alone (or nearly alone) control a nation's policies on national security and war? Or are there practical alternatives?

In his book, *Exercise of Power*, Robert Gates calls for "The Symphony of Powers," by which he means that nations should use a panoply of methods to exercise power, not just its military. A comprehensive array would include at least these seven methods: economic measures, strategic communications, relevant technology, diplomacy, foreign assistance, ideology, cultural outreach, and covert operations. Gates asserts:

> *Throughout history, power has most commonly been defined in terms of the ability to coerce obedience or submission by force of arms. But it is a mistake to think of power only in those terms ... I argued as secretary of defence that the American government had become too reliant on the use of military power to defend and extend our interest*

*internationally, the use of force had become a first choice
rather than a last resort.*

"First choice, rather than a last resort." The idea cries out for explication. War is horrific, costly in lives and treasure, and can damage the standing of leaders who order it. Why prefer it to other means of exercising power? The answer must come from a factor already discussed: Human males have an affinity for combat and war – not an instinct or compulsion, but a strong inclination to address disputes with violence. War has the appeal of seeming direct, decisive, fast. There's no "pussy-footing around," no waiting for non-military means to take effect. Leaders can appear, at least for a while, strong and dominant (Remember President Bush's "Mission Accomplished" slogan?) This kind of posturing often wins approval from other men and the admiration of women. Another result is perpetual war and a tragic failure to achieve a nation's constructive goals.

Compounding the problem, some nations have a history of using false evidence to justify war. A prominent example: According to an article published by the U.S. State Department's Office of the Historian, the United States government in the early nineteen-sixties opposed an election in South Vietnam that some male officials thought would unite the country under communist rule. This belief led to sending U.S. warships to the Gulf of Tonkin, off the coast of North Vietnam. On August 2, 1964, North Vietnamese patrol boats fired (ineffectually) on an American warship in the Gulf of Tonkin. The United States then claimed that a "second attack" occurred when North Vietnamese boats allegedly fired torpedoes at an American destroyer on August 4, 1964. President Johnson immediately asked Congress for war powers and escalated operations against North Vietnam.

That decision could be seen as timely, decisive action against a dangerous enemy. But an article on history.net by a retired U.S. Navy captain states that the alleged second attack never happened: Navy Commander James Stockdale (later Ross Perot's running mate) was flying over the area when the alleged attacks were supposed to have occurred. He said, "Our destroyers were just shooting at phantom targets ... There were no [North Vietnamese] boats there." The captain of one of the U.S. destroyers attributed the initial reports from his crew to "over-eager sonar operators." This view of the incident was later supported by the former Defence Secretary, Robert McNamara, the National Security Agency, and the Pentagon Papers.

The United States may have had other reasons to regard North Vietnam as an enemy, but if people in the Federal administration had taken more time to analyze and debate the total situation, they may have thought better of launching a war that eventually cost more than 58,000 American lives and a million other lives, not to mention more than $1 trillion in today's dollars, according to thebalance.com. Instead of thorough investigation and appraisal, we had the age-old male reaction: A man with authority must not appear hesitant. He must not appear timid. He must, at any cost, "act like a man." That command, impelled by biological inheritance and a male-dominated society, often results in actions radically different from those of a responsible adult.

The U.S. invasion of Iraq in 2003 was also predicated on false evidence. As reported by UN inspector Hans Blix before the invasion, there were no weapons of mass destruction – a conclusion reached after 700 inspections. In addition, it is well-established there never were any "yellow cakes" from Niger and no "aluminum tubes" suitable for use in rockets or centrifuges. Apparently, these facts were either ignored, or not adequately investigated, before deciding to invade. Yet, according to the watson.brown.edu site, men in the United States launched a war that lasted 20 years, claimed roughly 300,000 lives, and cost trillions of dollars.

## Women equipped to avoid war

Would a female President have decided not to invade? Would a much larger group of female Congress members have asked more questions and debated the issues more thoroughly? One way to answer such questions is to realize that women are better suited to avoid war than men. For at least six reasons:

### 1. Men are inclined to violence, women are not

As described earlier, human males obtained food by hunting and killing large animals with crude weapons for nearly 300,000 years. This activity required a hormonal balance favoring violent aggression, as well as physical abilities that enabled tracking, running, and lethal spearing. Given the high survival value of the necessary genes, it's likely they were conserved by natural selection: Generation after generation, the most successful food providers probably fathered more children than did less successful hunters, so their genes have lived on.

In *The Most Dangerous Animal*, author and philosopher David Smith points out one effect of male inheritance by quoting Sigmund Freud:

"Men are not gentle, friendly creatures wishing for love, who simply defend themselves if they are attacked, but that a powerful measure of desire for aggression has to be reckoned as part of their instinctual endowment."

### 2. Women oriented to peace

In contrast, ancestral human females probably spent much of their adult life pregnant, nursing, or otherwise taking care of children, so their main activities would have been nurturing and gathering edible plants and fruit. This also applies to grandmothers, who were probably an important part of keeping the local group alive and well – they knew more about infant care than young mothers, and more about which foods to eat and which to avoid. These abilities would have been genetically conserved; groups that were good at nurturing and gathering probably brought more children to sexual maturity than those that were not.

As mentioned earlier, women had to be assertive to get a fair share of food and comfortable living space, but they could not do it physically; the men were bigger, stronger, and more violent. So women had to achieve their goals by debate and negotiation – useful abilities when strategizing how to avoid a war.

### 3. A peace-oriented brain

Given the substantial difference in the ancestral lifestyle of men and women – as well as the probable effect of natural selection – it would be reasonable to expect the female brain to be different from the male brain. Yet for many years, it was thought that any differences were minor and unimportant. In a 2017 article in *Stanford Medicine*, Bruce Goldman reports that view has changed:

> *But over the past 15 years or so, there's been a sea change as new technologies have generated a growing pile of evidence that there are inherent differences in how men's and women's brains are wired and how they work.*

Many studies support this finding as a key characteristic of men and women: An article on Columbia.edu in 2020 states, "The parts [of the brain] that women used to perform their responsibilities increased in size while the parts that the men used for their activities (such as hunting) became larger compared to the female counterparts." Also, a 2014 article by Uphadhayya and Guragain in the Journal of Clinical & Diagnostic Research states,

*Male and female brains show anatomical, functional and bi-*
*ochemical differences throughout life. Women, for example,*
*have a larger brain area devoted to verbal communication*
*and interpreting non-verbal cues.*

These abilities are obviously useful in raising children – or in con-
ducting effective diplomacy.

### 4. A peace-oriented hormonal balance

The Stanford Medicine article mentioned above attributes brain differences to a difference in hormonal balance:

*But why are men's and women's brains different? One big*
*reason is that, for much of their lifetimes, women and men*
*have different fuel additives running through their tanks: the*
*sex-steroid hormones ... Importantly, males developing nor-*
*mally in utero get hit with a big mid-gestation surge of tes-*
*tosterone, permanently shaping not only their body parts*
*and proportions but also their brains.*

In a 2018 issue of *Endocrine Reviews*, David Handelsman and colleagues quantify the difference in hormonal balance. The authors report that the far higher levels of testosterone in men, versus those in
women, can have a significant effect on their potential for peak athletic performance:

*From male puberty onward, the sex difference in athletic*
*performance emerges as circulating testosterone concentra-*
*tions rise as the testes produce 30 times more testosterone*
*than before puberty, resulting in men having 15- to 20-fold*
*greater circulating testosterone than children or women at*
*any age.*

This finding becomes even more significant when we remember that testosterone aggravates any tendency toward aggressive behavior that may already be present. With a far lower level of testosterone and other androgens, women are better equipped than men to conduct a dispassionate discussion about the prospect of going to war.

## 5. Peace-oriented type of aggression

There's no doubt that women can be just as aggressive as men. But women tend to have their own way of showing it, as reported by Kaj Bjorkqvist, a professor of developmental psychology at a Finnish university. In a psychology journal article published in 2018, he states:

> *Studies on gender differences in aggressive behavior are examined. In proportions of their total aggression scores, boys and girls are verbally about equally aggressive, while boys are more physically and girls more indirectly aggressive.*

In a 2021 article in *Current Anthropology*, Lise Eliot, a Chicago Medical School professor, reports a similar finding:

> *Of the various behavioral differences between males and females, physical aggression is one of the largest. Regardless of gender, children's physical aggressiveness peaks between two and four years of age but then starts diverging, as girls learn more quickly than boys to suppress such overt behaviors. By puberty there is a sizable gender difference in physical aggression and violence.*

Studies like these indicate that women are far from the docile, submissive creatures some men would like them to be. At the same time, women tend to avoid expressing anger and aggression in physical terms, like those implicit in waging war.

## *6. Less violent criminal behavior*

We have just reviewed some of the reasons to expect that women will behave differently from men in dealing with disputes. But does that difference carry through to the strong emotions implicit in violent criminal behavior? Data on violent crime indicate a positive answer. Based on a 2007 study by the U.S. Department of Justice, the HowStuffWorks website tells us men committed 75.6 percent of violent crimes, while women committed only 20.1 percent. (The gender of the remaining 4.3 percent could not be determined.)

The National Center for Biotechnology Information provides additional evidence. After analyzing FBI data on U.S. homicides between 1976 and 1987, the Center reports: "Although women comprise more than half the U.S. population, they committed only 14.7% of the homicides noted during the study interval." Comparing the percentages, U.S. men in the study murdered nearly six times more often than women. Worldwide, according to a 2013 report by the United Nations Office on Drugs and Crime, 96 percent of all homicide perpetrators were men. So even when committing a criminal act, women are far less violent than men, a trait that would help to avoid needless military action.

These facts show that women are substantially less violent than men and far more likely to debate and negotiate before resorting to military action. At the same time, history shows that women will go to war in certain situations. We are not talking about a day-and-night difference, but rather a difference that seems likely to lessen the chances of needless wars, such as the U.S. invasion of Iraq.

## *7. Better results in negotiation*

Women's tendency to debate and negotiate can have benefits beyond avoiding needless wars. Writing in aeon.org, science journalist Josie

Glausiusz provides an example of women's superior negotiating skills in her report on the findings of Mary Caprioli, a University of Minnesota professor:

> *Caprioli's data show that, as the number of women in parliament increases by 5 percent, a state is five times less likely to use violence when confronted with an international crisis (perhaps because women are more likely to use a 'collective or consensual approach' to conflict resolution) ... an analysis by the U.S. non-profit Inclusive Security [examined] 182 signed peace agreements between 1989 and 2011 and found that an agreement is 35 per- cent more likely to last at least 15 years if women are included as negotiators, mediators and signatories.*

This point brings us back to the national security policy of Robert Gates, as outlined earlier. He advises the U.S. (and other nations) to employ a broad array of methods before resorting to military action, which, he says, should be a "last resort." Based on the differences in how men and women tend to deal with disputes, women may be more inclined to follow Gates's advice than men – and that is likely to make the world a more peaceful place.

Returning to our origins, it was women who had the job of keeping vulnerable infants and children alive in dangerous environments. In addition, each human mother had that job for at least a decade for each child, because human children develop slowly. It follows that women had a strong interest in keeping their local environment peaceful. Yelling and aggressive body postures in the home area may have been tolerable; physical harm was not. It's no surprise, then, that natural selection would favor females averse to violence. In contrast, men obtained food for their groups by frequent lethal violence.

Today, it's clear that neither sex has a monopoly on constructive judgment. There may be times when a nation has to respond to an attack with speed and military action. There are many other times when caution and non-violent methods are more productive. Japan in the nineteen-seventies, for example, found it could accomplish far more with cars and television sets than it could with bombs and bullets in the nineteen-forties. Economic and technological power can often be more useful than military power.

## A call for balance

Men have been almost completely in charge of deciding to wage war for thousands of years, and the results have been catastrophic. It is time – past time – for women to have an equal voice in deciding when to resort to military action. Women have shown they can be effective military officers in executing war policies; now it's time for women to play a leading role in creating war policies.

> Specifically, if there are 10 people sitting around a table deciding whether or not to go to war, at least five of them should be women.

There's another component. A woman's inclination to question and debate probably depends, to some extent, on the number of women in her organization. If only a small percentage of a governing body consists of women, those women (like other minorities) will likely feel peer pressure to vote as the men do, as well as pressure from voters who believe that women with authority should act like men.

To get the full benefit of women's potential contribution to rational policies, a country needs a high percentage of females at all levels of government – federal, state, and local. This is how we can change national attitudes about the appropriate role of women in governments. It may also be our best hope for more constructive defense, economic, environmental, and social policies.

At least two objections immediately arise: When women have been heads of state, some have not avoided war; Margaret Thatcher and Indira Gandhi come to mind. One answer to this objection is that women respond to situations. If a nation is attacked, it goes to war. It doesn't matter who is president or premier. Case in point: Israel's female premier presided over several wars. But Israel was fighting for its survival; it would have gone to war if the premier had been a goldfish, instead of Golda Meir.

Another objection is that many nations, especially those in the Mideast and some in the Far East, would never consider giving women a powerful voice in national affairs. We would then be faced with an asymmetrical situation: A potential adversary would be playing by different rules and might try to take advantage of a nation making a serious effort to avoid war. But history shows that women leaders will indeed prepare for war and go to war under certain circumstances. There is no evidence that female heads of state, when pushed, will back down any more than men would in the same situation. That said, we have reviewed reasons to believe that women in power would help a country avoid, if not all wars, then some wars, which, for the United States and some other countries, would be a major change for the better.

## A gender-centered obstacle

Despite some progress, women are still hobbled by certain attitudes held by many men (and some women): One could be called the "Can't Win Catch:" When a woman asserts herself, she is often cast as unfeminine or "bitchy." If she acts deferentially, she is dismissed as a "lightweight" not fit for high office. To progress in any field, she often has to tip-toe between those two notions, a maneuver that men are free to avoid. It's no surprise that these attitudes also apply in politics. An October 2020, *New York Times* article states:

> *Research has found that it is much harder for female candidates to be rated as 'likable' than men — and that they are disproportionately punished for traits voters accept in male politicians, including ambition and aggression. At the same time, voters view their credentials more skeptically and question their toughness, a precarious situation that is so universal for women seeking leadership roles that it is known as the 'double bind.'*

Another destructive attitude goes something like this: "It's good to be aggressive. You need aggression to make tough decisions and get things done. Men are naturally more aggressive than women. So if you want something done, give the job to a man. Of course, some women can be good managers, but mostly when they act like men."

Most people in the West don't say that anymore. But many think it, and the fallacy is still operative. It helps to explain, for example, why a female legislator might feel compelled to vote for a war she doesn't really support: In a male-dominated setting, she has to show she can "act like a man." But is the general premise about the efficacy of aggression valid? Does sheer aggressiveness add validity to a decision? If men are naturally more aggressive than women, does that mean they're necessarily better at running companies – or countries? Let's consider some possibilities:

What if male aggression now operates all too often as an atavism, a counter-productive relic of our distant past? What if fundamentalist religions demand strict adherence to doctrine and dogma, not to please God, but to bolster male authority? What if, in fact, excessive male aggression is a primary source of needless wars and destructive practices in government, business, science, and other fields?

These possibilities and probabilities indicate that it's time for women to have at least as much political power as men.

## Women in the lead

The issue of whether women can lead has long been settled. In Western countries, for example, the most recent notable event may be the election of Kamala Harris as Vice-President of the United States. Several U.S. Presidential candidates in 2020 were women, and a woman – let us not forget – was America's popular-vote choice for President in 2016 by nearly three million votes. The U.S. had a woman (Janet Yellen) as the head of its Federal Reserve System, which has a major role in managing a $27 trillion economy; in 2021 Yellen became the U.S. Secretary of the Treasury. As of 2020, U.S. lawmakers included 25 female Senators and more than 100 female Congressional Representatives.

Elsewhere, women may play an even larger role in government. Germany, the most influential country in Europe, was led by Angela Merkel (a former physicist) for 15 years. A Council on Foreign Relations article shows that 21 countries had a female head of state in 2020.

## Female leaders: better at managing Covid-19

An article by Amanda Taub in the April 24, 2020, *New York Times* reports that four countries with minimal death rates from Covid-19 (Finland, Germany, New Zealand, and Taiwan – are all governed by women. Each took a risk-averse approach to dealing with Covid by establishing strict programs of testing and social distancing soon after the virus appeared. In contrast, governments taking a bold approach – the United States and the United Kingdom – were governed by notably aggressive males and had high death rates. Taub's article points out the potential merits of what some might call a female style of governing:

> *That style of leadership may become increasingly valuable.*
> *As the consequences of climate change escalate, there will*
> *likely be more crises arising out of extreme weather and*
> *other natural disasters. Hurricanes and forest fires cannot*
> *be intimidated into surrender any more than the virus can.*

Turning to business, we find that only about 40 Fortune 500 companies have female CFOs – not nearly enough, but the companies include heavy-weights like General Motors, IBM, PepsiCo, Lockheed Martin, Oracle, General Dynamics, and Citigroup. (A few examples may no longer be current.) This group of female CEOs indicates that gender is no bar to leading huge organizations and that greater representation in this area awaits only society's recognition of female capability. In addition, women manage tens of thousands of smaller companies and other kinds of organizations throughout the U.S. The 2022 annual report of the National Women's Business Council says that women-owned 10.9 million (41%) of non-employee U.S. firms and 1.2 million (21%) of employer businesses.


## Women worldwide

It's not necessary to cite women as heads of state or as Fortune 500 CEOs to see their capabilities in many fields. A website from UN Women (beijing20.unwomen.org) provides inspirational biographical sketches of more than 40 women in at least 34 countries. The breadth of their achievements is impressive by any standard: Solo transatlantic sailor. Solo skier to the South Pole. Military and civilian pilots. Holders of governing positions even in countries with daunting obstacles to women. Professor of medicine. Medical researcher. Inventor of computer technology. Mine-clearer. Veteran climber who has summited the seven highest peaks in the world. Cliff parachute-jumper. Deep-sea scuba diver. Military officer. NASA scientist. One could argue this list and others like it should be required reading for children in elementary and secondary schools.

In October 2020, came an event that points to an even larger role for women in science: French microbiologist Emmanuelle Charpentier and American biochemist Jennifer Doudna were awarded the Nobel Prize in Chemistry for discovering the CRISPR/Cas 9 gene editing system, a low-cost way of editing genes to produce desirable genetic changes, one type of which may permit curing inherited diseases.

What the world needs is more of the same. More women achievers in diverse fields. More women with executive power. More women voting on fundamental issues. More women helping to make societies less violent, more equitable, more humane.

**War not inevitable**

War only looks inevitable, because societies have allowed a single sex to control how government responds to disputes, a policy that ignores the potential contributions of half of humanity. That has to end. Societies everywhere will be more constructive, economies more robust, and war less frequent when women's voices and political power are at least equal to those of men. An important complement to this change is Robert Gates's call for governments to use a multifaceted approach to manage national security issues. That approach – wielded by women and men working together – offers the world's best chance of preventing needless wars and developing constructive societies.

# 13

## Blueprint for a Sane Society

In previous chapters, we established several fundamental facts that support a national goal of women assuming political power equal to that of men. This strategic change has become more urgent because of the increasing threat of global war and the planet-wrecking effect of global warming. The world needs a similar change in the gender profile of corporate management as an end in itself and to help achieve gender parity in local, state, and federal governments.

### Background

As discussed, both chimpanzee males and human males inherit a propensity for combat and war, which is why men have been at war almost continuously for thousands of years. This propensity comes as one effect of natural selection operating over nearly 300,000 years, during which male humans earned a living by killing large animals for food. This activity required not only certain physical attributes – the ability to run, sweat, and throw spears – but also a plentiful supply of testosterone and other androgens, to help men confront the danger of flailing horns and hooves.

Women had multi-faceted tasks requiring a wide range of abilities, including bearing new life, nursing, nurturing, and safeguarding small children. They probably also foraged for edible plants and hunted small animals. Judging by the available evidence, it appears that natural selection operated to reinforce those capabilities, as well.

The result of these parallel histories is complex. Women tend to be better than men at some kinds of activities; men tend to be better at others. Overall, available evidence indicates the two genders are equal in their ability to carry out the kinds of tasks found in post-industrial societies. They appear to be equal in their abilities to lead a country or manage a company. An individual man may excel in activities requiring some form of aggression. A woman may excel in activities that call for skill in interpersonal relations. Since the typical management job demands a mixture of both, performance usually depends on how well a person's skill set matches the requirements of the job, regardless of gender.

## A difference to be nourished

Despite historical exceptions, women seem far less willing to wage war than men. As reported in earlier chapters, they are more inclined to discuss and debate divisive issues than to brandish missiles and bombs. This difference is a major asset at a time when atomic scientists, prompted by new kinds of weapons and other threats, have set the Doomsday Clock closer to apocalypse than ever before. Women seem particularly well suited to implement the broad array of statecraft measures advised as alternatives to war by former Secretary of Defence, Robert Gates, in his 2020 book, *Exercise of Power.*

But women as a group cannot step into a greatly expanded role in government or companies without adequate preparation. Enabling women to gain political and corporate power equal to that of men calls for a massive effort from many parts of society, including universities, parents, government, religions, businesses, and motivated individuals. In this chapter, we'll review some of what needs to be done and the resources currently available. We begin with several observations:

Every nation needs to make a conscious decision about the message of this and similar books: Are we serious about wanting women to have a role equal to that of men in government and society? If the answer is 'yes,' then society must help women achieve that role. An immediate objection might be, "We don't help men because of their gender. Why should we help women? Sounds like discrimination."

The answer is that every society still relies on women to do the most important job in any society: give birth to and nurture each future generation. Making this contribution requires a huge investment of time, energy, and share-of-mind. All too often, it also means giving up a career in government, business, science, sports, or the arts. Society therefore owes women something in return. It owes them affordable, effective solutions to some of the goals and obstacles that women must confront.

## Women's practical challenges

The issues come in various shapes and sizes, some psychological, like a societal belief that women are somehow less capable than men, and others, hard-edged and practical, like these:

- If I decide to go back to work before my child is ready for preschool, how do I find affordable and reliable childcare?
- My child is ready for preschool. Is there a good one near me that I can afford?
- I just had a baby and want to stay home for at least two months. But we can't afford it.
- Will employers be willing to hire and promote me, knowing that I'm a mom?
- I want to breast-feed my baby. Will my employer allow it?

- Sex is part of family life. How do I get affordable and accessible pregnancy control?

## The solutions women need

With these challenges in mind, a society needs at least seven advances in public policy to help more women achieve positions of political and corporate power:

- Affordable, high-quality childcare.
- Many more affordable pre-schools.
- Paid family leave.
- Affordable pregnancy control.
- Employers that hire and promote mothers.
- A practical way to learn marketable skills at home.
- Employer flexibility about breast feeding.

The federal Break Time for Nursing Mothers law requires employers covered by the Fair Labor Standards Act (FLSA) to provide basic accommodations for breastfeeding mothers at work. But the U.S. lags most European and some Asian nations in the other initiatives. The sooner we make a serious effort to catch up, the better. What's at stake is not only reducing the threat of needless, destructive wars, but also creating a healthier, more equitable, more humane society. Let's take a closer look at six of the proposed policies mentioned above.

## We need affordable, high-quality childcare

Childcare.gov points out that childcare is one of the biggest items in family monthly budgets – often more than the cost of housing, college tuition, transportation, or food. Their site lists a variety of ways to help, including financial assistance; work- and school-related programs;

tax credits; and special programs for Native Hawaiians, Alaskans, and Native Americans.

In 2019, the Center for American Progress, a non-partisan policy institute, published this report on their website:

*Whether due to high cost, limited availability, or inconvenient program hours, childcare challenges are driving parents out of the workforce at an alarming rate. In fact, in 2016 alone, an estimated 2 million parents made career sacrifices due to problems with childcare...American business, meanwhile, loses an estimated $12.7 billion annually because of their employees' childcare challenges.*

The same article points out there are two federal programs with free or subsidized childcare for low-income families: (1) Child Care and Development Block Grant and (2) Head Start. Both have severe limitations. Only 15 percent of eligible families are able to receive subsidies through the block grant program and, in most cases, the subsidy is too low to support high-quality childcare. Head Start does deliver high-quality early education, and – what is often overlooked – comprehensive health and social services to one million low-income children. But that one million is only one-third of the eligible 3- to 5-year-olds. Early Head Start (children under three) serves only seven percent of eligible children.

## We need affordable preschools.

America needs comprehensive, affordable, preschool and day care facilities nationwide – if not just out of compassion, then because it pays. The National Institute for Early Education Research reports a huge return on investment (ROI) from preschool education. Here's a quote from their website:

*Although high-quality preschool requires a sizable invest-
ment, the evidence suggests the returns at local, state, and
national levels outweigh the costs. While any return that ex-
ceeds $1 for every $1 spent indicates that a program pays
for itself, cost-benefit analyses show high-quality preschool
programs can yield up to a $17 return for each dollar in-
vested, when lifetime outcomes that result in contributions
to society are considered.*

A 17-to-1 ROI sounds like a powerful motivation to make the necessary budget allocations. It's important to be careful with taxpayer
money and to avoid needless debt. But affordable preschool education is not just a case of what some might call "more government
spending;" it's a vital investment in the prosperous future of any society.

### We need paid family leave

The U.S. should make it possible for working mothers to have home
leave after they've given birth or in case of special needs. Programs
that soften the financial strain of being a mother are vital to the health
and well-being of children and the productivity of women.

In 2015, the *Huffington Post* reported how other nations were
handling family leave:

*In Sweden, parents are given 480 paid leave days per child
which can be used between moms and dads, and many
other European countries are not far behind. Spain offers
112 paid days, the UK offers 280 days with 90 percent pay,
France offers 112 paid days, and Italy offers 140 days with
80 percent pay.*

What about the U.S? America has no national policy of paid leave, and even unpaid leave is difficult to get. The Huffington Post article reveals that "fewer than one-half of the nation's private sector workers are eligible for leave." In addition to the U.S., only three other countries do not mandate paid leave for mothers: Lesotho, Swaziland, and Papua New Guinea. These facts prompt a question: Why doesn't the world's richest country have family leave policies that at least match those of other advanced nations?

## We need affordable pregnancy control

To become a more prominent voice in national and international affairs, women need easy access to practical methods of preventing unwanted conception. Deciding whether or when to have a child is perhaps the most important decision a woman can make. This decision should be as planned and deliberate as a couple can possibly achieve. Also implied is that governments should do all they can to help couples reach this goal.

In the U.S., there was broad support for pregnancy control until October 2017, when the federal government made it possible for any employer to get a religious exemption to insurance coverage for pregnancy control. A 2017 article on vox.com states:

> *Worldwide, the ability to plan pregnancy is associated with lower infant and maternal mortality, lower mother-to-child transmission of HIV, and fewer abortions–especially unsafe abortions...In one survey of patients at family planning clinics, 64 percent said birth control helped them extend education, 71 percent said it helped them support themselves financially, and 77 percent said it helped them take care of themselves or their families.*

The vital benefits of planned conceptions bring up an issue: Should there be no limit to the acceptance of a religious dogma imposed by a minority, even when its application harms society as a whole? The question is especially pertinent when we consider the actual practice of those who profess such beliefs. It is well-known and documented that most people who follow a religion opposed to birth control do not adhere to that aspect of their faith. The Guttmacher Institute reports the use of contraceptive methods other than "natural" by "99% of all sexually experienced women," including "98% of those who identify themselves as Catholic." With these facts in mind, how is there a net gain in freedom if a minority can deny the majority a benefit permitted by their country's elected government?

These points are especially relevant since the U.S. Supreme Court overturned Roe vs. Wade. Six of the nine justices are Catholic.

## We need employers who hire and promote women

Employers know there's a potential downside to hiring young women: they may quit to become full-time moms; or nurturing their children may at times take precedence over their work. So be it. This risk is a necessary cost of helping women achieve a larger role in business and government, a process that has the potential to help any nation avoid needless wars and achieve a better quality of life for everyone. The policy could be implemented by a system of tax incentives for organizations that hire a stipulated percentage of young women.

## Women need a way to learn marketable skills

There are plenty of ways to take online classes designed to duplicate an academic school program. What women also need are ways of gaining or improving marketable skills.

An interesting example is Inc's online article, "14 In-Demand Skills You Can Learn Online Now." For each skill, the site provides links to sources offering online instruction. Of course, the instruction comes at a cost, which not everyone can afford. So there also needs to be a system that offers at least a partial subsidy to help with tuition. Tax incentives could be offered to companies that step up and offer such assistance; the companies could then also benefit from having a stream of future job applicants with known qualifications. State and federal education budgets could be expanded to provide additional support.

**Seeking sanity**

Together, these policies constitute what could be called a "Blueprint for a Sane Society." Why "sane?" Because the United States and other prominent nations are engaged in the obsessive development of increasingly dangerous and costly weapons, while global warming promotes severe flooding, massive fires, crop loss, deadly pollution, and perhaps, irreparable damage to the planet – all of it self-inflicted. This behavior can be considered not merely insane, but criminally insane.

The total cost of the Blueprint would be high, but so would the rewards in terms of a more equitable and productive society. Not the least reward would be getting more women into government, with the well-grounded hope of avoiding needless wars. Also, a higher percentage of qualified women in government might improve the cost-effectiveness of military spending. With far less interest in combat, women are less likely to be swayed by the sex appeal of technically exciting weapons that offer no real gain in national security. To manage the Blueprint's high cost, its policies could be implemented over several fiscal years, perhaps extending to a decade.

**Objection and Response**

Objections to the Blueprint for a Sane Society will most likely be ideological and financial. Some people will say, "We can't do that – it would be socialism." The answer is simple: Do we want to live by labels and dogma? Or by doing what is known to improve the lives and health of a country's citizens?

The financial objection can be summed up by the claim, "We can't afford it." Let's have a look at that: For fiscal year 2022 the U.S. military budget was $877 billion. As mentioned, that's 10 times the amount spent by Russia and three times the amount spent by China, according to *Axios*. Another way of looking at our military spending: It's more than the spending of the next 10 countries *combined*.

A case can be made for taking a mere 20 percent of that amount – $175 billion – and spending it to nurture and educate our children, while helping women achieve a much larger role in society and government. As we have seen, spending on advanced weapons development tends to function as a zero-sum game that does not necessarily improve national security. We need to remember President Eisenhower's warning about the dangers of military spending:

> *In the councils of government, we must guard against the acquisition of unwarranted influence, whether sought or unsought, by the military-industrial complex. The potential for the disastrous rise of misplaced power exists and will persist.*

Eisenhower spoke those words in 1961. In the light of planned nuclear modernization estimated to cost over $1.2 trillion over 30 years, we need to ask, "How much of that is required for national security? And how much of it exemplifies Eisenhower's warning?"

A case can be made that a large chunk of the Pentagon's budget can be cut without impairing national security. A recent example: The Pentagon's 2021 plans call for spending $100 billion to produce a new nuclear weapon. An article in the February 24 issue of the

*Bulletin of the Atomic Scientists* raises serious questions about the need for such a weapon, suggesting the $100 billion can be spent in more useful ways.

Another source of income to fund the Blueprint program would be to levy higher taxes on large corporations and high-net-worth individuals. History indicates this can be done without hurting the economy. According to the nonpartisan Congressional Research Service, changes in the top tax rate "do not appear to be correlated with economic growth." Example: When Presidents Clinton and the elder Bush raised taxes, the increases were followed by sharp gains in GDP. When the younger Bush cut taxes, the growth rate sank and stayed at low levels for several years (source: Bureau of Economic Analysis and Haver Analytics). As some legislators and economists have suggested, a tax on total wealth should also be considered.

A microcosm of affordability comes from the Oregon county that includes Portland. In 2020 the county launched a program that will provide free preschool for every three- or four-year-old child in the county. It will be paid for by a progressive tax on high earners. Overall, it appears that the United States can afford the same kinds of programs that Europeans and some Asian countries have had for decades.

It will take years for elements of the Blueprint to become widely available. In the meantime, girls and women can turn to a wide variety of organizations to help raise their profile in government and society.

## ORGANIZATIONS FOR GIRLS

Any serious national effort to bring more women into positions of power should start with girls. Few people can jump from being a so-called "sweet young thing" to stepping out front and saying, "I'm here to lead." Fortunately, the U.S. has lots of leadership programs for sub-teen and teen-aged girls. Some are state-wide, some are geared to vulnerable young women, some aim to help girls interested in science and software engineering, and some cater to young adult women who are already on leadership tracks. Here are three examples:

*Girls Leadership*

One of this group's core beliefs is that leadership should extend beyond the few in formal roles, such as a team captain or class president. Girls Leadership shows girls how to turn everyday relationships with friends, family and peers into leadership opportunities. (girlsleadership.org)

*Girls Inc*

With a theme of "inspiring all girls to be strong, smart, and bold," this organization also states, "Our comprehensive approach addresses all aspects of a girl's life and helps her discover and develop her inherent strengths. Girls receive programming to grow up healthy, educated, and independent." (girlsinc.org)

*The Foundation for Girls*

This group's mission asserts that "the journey to self-leadership, economic and social mobility is tied to a fundamental understanding of our core values, key character traits, and how we fit into the larger world we live in ... our program focuses on helping each girl understand who she is and what she's striving for." (foundationforgirls.org)

## ORGANIZATIONS FOR WOMEN

There are many organizations dedicated to helping women get elected to public office, or to advance the role of women in our society in other ways. Here are just a few with specific political goals, listed in alphabetical order:

*CAWP*

One of the most prominent organizations fostering female leadership is the Eagleton Institute of Politics at Rutgers, the State University of New Jersey, and home to the Center for American Women in Politics (CAWP). Their mission is to "promote greater knowledge and understanding about the role of women in American politics, enhance women's influence in public life, and expand the diversity of women in politics and government."

In 2013, CAWP helped convene the White House Conference on Girls' Leadership and Civic Education. The Center's report on that conference and its results are now available at tag@eagleton.rutgers.edu. One of the Center's programs is Teach a Girl™. It's designed for parents, educators, leaders of youth-serving groups, and media outlets for young audiences. By training teachers, the program leverages communication to spur positive change. A useful feature of the Center's website is a map with each state outlined. Click on a state and the viewer sees a list of leadership programs designed for girls and young women in that state. (tag.rutgers.edu/programs-places)

*Emily's List*

Describes its mission as: "We elect pro-choice Democratic women to office." A few measures of its size and impact include:

- A community of more than five million.
- Over 1,200 election victories.
- 9,000 women trained.
- More than $600,000+ raised for candidates.

Emily's List describes part of its vision in these terms: "We will work for larger leadership roles for pro-choice Democratic women in our legislative bodies and executive seats so that our families can benefit from the open-minded, productive contributions that women have consistently made in office." (emilyslist.org/pages/entry/our-mission)

Associated with Emily's List is the 2021 book, *Run to Win: Lessons in Leadership for Women Changing the World*, by Stephanie Schriock, former president of Emily's List, and co-author Christina Reynolds. Published by Dutton, the book is a manual showing how women can do what the title says they can do. Foreword by Vice President Kamala Harris.

## *League of Women Voters*

Its aims are: "Empowering voters. Defending democracy." Its stated vision: "We envision a democracy where every person has the desire, the right, the knowledge, and the confidence to participate." The League works on a national level and supports "over 800 state and local Leagues in priority issues with the Campaign for Making Democracy Work." (lwv.org)

## *National Foundation for Women Legislators*

This organization's stated purpose is to provide strategic resources to elected women, an exchange of diverse ideas, and effective governance through conferences, state outreach, education materials, professional and personal relationships, and networking. (womenlegislators.org)

## *National Organization for Women (NOW)*

With more than 500,000 members, NOW ranks as one of the largest women's political organizations in the U.S. Its website states:

"NOW is a multi-issue, multi-strategy organization that takes a holistic approach to women's rights. Our priorities are winning economic equality and securing it with an amendment to the U.S. Constitution that will guarantee equal rights for women; championing abortion rights; reproductive freedom and other women's health issues; opposing racism; fighting bigotry against the LGBTQIA community and ending violence against women...NOW has hundreds of chapters and hundreds of thousands of members and activists in all 50 states and the District of Columbia." (now.org) A NOW Political Action Committee supports female candidates financially. (now-pac.org)

*National Women's Political Caucus*

Its mission statement emphasizes financial support and training: "a multi-partisan grassroots organization dedicated to increasing women's participation in the political process. NWPC recruits, trains, and supports pro-choice women candidates for elected and appointed offices at all levels of government. In addition to financial donations, the Caucus offers campaign training for candidates and campaign managers, as well as technical assistance and advice." (nwpc.org)

*Running Start*

Describes itself as "empowering young women to get involved in politics and transform our world, one elected female leader at a time. We offer programs that equip them with the hands-on drills and confidence they need to run and win." A non-partisan organization, Running Start in 2023 has more than 100 training programs nationwide and has so far trained more than 15,000 women. (runningstart.org)

*She Should Run*

---

This group affirms diversity: "By identifying and tackling the barriers to elected leadership, SSR convinces women from all political leanings, ethnicities, sexual identities, and backgrounds to see themselves as future candidates. Our programs unveil the many pathways to leadership, guide them toward discovering 'why,' and connect them with a supportive community." (sheshouldrun.org).

*Women's Campaign Fund*

---

Calls for an equal role for men and women in government, stating: "a national nonpartisan organization that commits to 50/50 representation by women and men in elected offices nationwide by 2028. We are people from all political parties who believe government works best when America is represented by 100% of the available talent, wisdom, and skill – 50/50 men and women, like the population." WCF supports women who have the "personal, professional, and political capabilities required to win office and govern effectively." (Type "women's campaign fund" in browser.)

**General welfare**

Organizations focusing more on the general welfare of women include: the Feminist Majority Foundation; the Guttmacher Institute; the National Partnership for Women & Families; the Public Leadership Education Network; Wider Opportunities for Women; and Women Strike for Peace.

This is only a partial listing of the productive organizations that need and deserve support from women and men who believe that women should have a much larger voice in local, state, and national governance. When it happens, we all can expect better government at home and less war.

## A family orientation

One illustration of the possibilities of increased woman power comes from the rising number and percentage of women in Congress. In 2013, 19.1 percent of federal Congressional members were women; by 2019 the proportion had risen to 23.7 percent, and 26 of those women were mothers. This is a welcome increase, but for any country to get the full effect of how women can improve society, the percentage of women legislators needs to be much higher. It's not that women are brighter than men; it's that they tend to see life from a different perspective – one that shows much more awareness of what mothers and families need.

Example: A 2019 *Washington Post* article points out that female legislators have proposed adding partial pay to the 1993 Family and Medical Leave Act (which currently offers only unpaid leave), tax credits for eldercare, and tighter gun control. Legislators who are mothers also introduce more child-related bills than women without children, such as bills that call for easier access to early childhood education, affordable higher education, and access to breast-feeding facilities.

# Vote for Women

This book shows that men, in general, have an abiding propensity for violence and combat, which often leads nations into needless wars. Women, in general, do not have this inclination. It therefore makes sense to form governments where women wield at least as much political power as men. In practice, this means that if 10 people are sitting at a table to decide if their country should go to war, at least five should be women. If the reader agrees with this general principle, it would make sense to vote for female candidates and donate to women's organizations. These actions may be the most practical route to a better society – and a more peaceful world.

# Epilogue

We've reviewed evidence indicating that human males have an atavistic propensity for combat and war. That doesn't apply to every man or in every instance, but it applies often enough to foment continual wars and the obsessive development of massively lethal weapons. We've also seen why this tendency is particularly dangerous in today's world, where authoritarian and terrorist regimes can get advanced weapons backed by artificial intelligence.

We've discussed why having more women in positions of political power, especially at the federal level, is likely to alleviate the threat of perpetual war. Finally, we explored the Blueprint for a Sane Society, which includes seven steps to help women advance in the political and corporate worlds. Perhaps the most basic point of *Why Men Make War* is that it suggests a practical and broadly beneficial way to reduce the number of needless armed conflicts.

Along the way, we reviewed evidence indicating how humans came to be who we are: We evolved to become a *Homo* species largely because our upright posture freed our hands to make and use tools and weapons, which helped to enlarge our brain and develop speech. It's likely that we became *Homo sapiens* mostly by our cooperative, 10-part Quest for Meat. Other evidence suggests that our ancestors cooperatively killed prey animals, proto-humans, and other humans seen as competitors. But at no point was human development the result of a single factor. We are, instead, the result of a serial interweaving of countless strands of biological, geographic, climatic, and cultural events.

# Additional Reading

For those who want to learn more about men's affinity for violence and war, here are several books that describe the condition in depth:

**Demonic Males**: *Apes and the Origins of Human Violence*. By Richard Wrangham and Dale Peterson, Houghton Mifflin, 1996. In this highly influential work, the authors reveal male chimpanzee behavior that helps to explain the propensity for violence and war that simmers in the nature of men. Along the way, they also show how eating meat and controlling fire led to a bigger and more powerful brain.

*T: The Story of Testosterone, the Hormone That Dominates and Divides Us*. By Carole Hooven. Henry Holt and Company, 2021. In the world of books written for both professionals and educated lay readers, this may be the definitive work on testosterone and its effects. A lecturer and department co-director at Harvard University, Hooven explains how testosterone dominates and divides the sexes. She also reports that much of the complex biology that leads to non-binary sexuality happens in the uterus, before birth. All this she does in conversational language that invites reading.

**Where Does Violence Come From?** *A Multidimensional Approach to Its Causes and Manifestations*. By Bernhard Bogerts, Springer Nature Switzerland AG, 2021. Professor Bernhard Bogerts, MD, is a neuroscientist and psychiatrist. Currently, director of the Salus Institute, which studies the causes of violence, Bogerts draws from an unusually diverse span of disciplines to explain how violence originates in our genes and in specific parts of our brain – and why an inclination toward violence is part of men's behavioral repertoire.

***Collapse****: How Societies Choose to Fail or Succeed*. By Jared Diamond, Penguin, 2006. Using historical examples, Diamond demonstrates the counter-intuitive fact that a nation governed by intelligent and educated people will not necessarily avoid self-destruction. More than 90 percent of the world's climate scientists have agreed that the human use of fossil fuels contributes heavily to global warming and its consequent damage to the planet. But the world is still taking baby remedial steps when giant steps are required.

***The Hunting Ape****: Meat Eating and the Origins of Human Behavior*. By Craig B. Stanford, Princeton University Press, 1999. Stanford is a professor of anthropology and co-director of the Jane Goodall Research Center at the University of California. No armchair expert, Stanford has spent years in the jungle, studying the behavior of chimpanzees. One of his main findings: Chimps place a high value on meat; sharing it with others is a key tactic to gain power and status within a group. Through natural selection, it may also be one of the traits that helped to enlarge the human brain.

***Exercise of Power****: American Failures, Successes, and a New Path Forward in the Post-Cold War World*. By Robert M. Gates, Alfred A. Knopf, 2020. After many years of public service, including a stint as Secretary of Defence, Gates advocates that nations use a broad variety of methods – a "Symphony of Powers" – to pursue their national interests. War remains a possibility, but only as a last resort. (*Why Men Make War* develops the thought that women may be better equipped than men to carry out Gates's advice.)

***How War Began****: By Keith Otterbein, Texas A&M University, 2004. Based on a thorough study of where, when, and how wars start, the book delivers on its title. Otterbein shows that violence and combative encounters have an ancient history and occur even in remote tribes.

**Constant Battles**: *Why We Fight*. By Steven LeBlanc (with Katherine R. Register), St. Martin's Press, 2004. LeBlanc reports that wherever he has worked in ancient sites, he has found clear evidence of violence and armed combat. He attributes war to growing populations depleting local resources.

**Sex and War**: *How Biology Explains Warfare and Terrorism and Offers a Path to a Safer World*. By Malcolm Potts and Thomas Hayden, BenBella Books, 2008. University of California professor Potts and Stanford University lecturer Hayden join forces to examine the relationship between male-on-male competition for mating opportunities and our inclination to violence and war. They point out that the more intelligent combatants were probably more successful in combat than the less brainy ones, which would have helped drive the evolution in the size of our brain.

**The Most Dangerous Animal**: *Human Nature and the Origins of War*. By David L. Smith, St. Martin's Press, 2007. A professor and director of the Institute for Cognitive Science at the University of New England, Smith makes the case that a penchant for group violence has been bred into us over millions of years of biological evolution. He explains how the skills and attitudes we developed to fend off predators came to be used to kill each other and cites many examples of faith-based brutality.

These 10 books, of course, represent only a fraction of the many written about why men go to war.

# Bibliography

Aiello, L. Wheeler, P. (1995, April 1). The expensive-tissue hypothesis: the brain and the digestive system in human and primate evolution. Current Anthropology, 36(2), 99-221. http://www.jstor.org/stable/2744104

Amadeo, K. (2020). Vietnam war facts, costs and timeline. The Balance. http://www.thebalance.com/vietnam-war-facts-definition-costs-and-timeline-4154921

American Psychological Association, Boys and Men Guidelines Group. 2018. APA guidelines for psychological practice with boys and men. http://www.apa.org

Ardrey, R. (1961). African genesis, New York. NY: Atheneum Books/Simon & Schuster.

Balter M., (2015 October 8). Homosexuality may be caused by chemical modifications to DNA. Science. http://www.sciencemag.org/news/ 2015/10/homosexuality

Belluck, P. (2019 August 29). Many genes influence same- sex sexuality, not a single 'gay' gene. The New York Times. http://www.nytimes.com/ 2019/08/29/science/gay-gene-sex.html

Berenbaum, S., Hines, M. (1992). Early androgens are related to childhood sex-typed toy preferences. Sage Journals, 3(3), 203-206. http://www.journals.sagepub.com/doi/abs/10.1111/j.1467-9280.1992.tb00028

Berenbaum, S.A. A spirited polemic takes aim at biological sex differences but misses opportunities to highlight relevant science. Science Mag. http://www.blogs.sciencemag.org/books/2017/01/18/723

Berlinski, D. (2019). Human nature. Seattle, WA: Discovery Institute Press.

Bogerts, B. (2021). Where Does Violence Come From? A Multidimensional Approach to Its Causes and Manifestations. Switzerland AG: Springer

Besant, A. (2012). Low levels of dopamine lead to aggressive behavior. PRI. http://www.pri.org/stories/2012-06-11

Bjorkqvist, K. (1994). Sex differences in covert aggression among adults. Aggressive Behavior, 20(1), 27-33. http://www.psycnet.apa.org/record/1994-23589-001

Bjorkqvist, K. (2018, February), Gender differences in aggression. Current Opinion in Psychology. 19, 39-42. https://pubmed.ncbi.nlm.nih.gov/29279220/

Blakemore, E. (2016, June). New evidence shows peppered moths changed color in sync with the industrial revolution. Smithsonian. http://www.smithsonianmag.com/smart-news/new-evidence-peppered-moths-changed-color-sync-industrial-revolution-180959282/

Bramble, D., Lieberman D. (2004, November 18). Endurance running and the evolution of Homo. Nature. 432. http://www.scholar.harvard.edu/files/dlieberman/files/2004e.pdf

Broad, W. (2021, January 24). How space became the next 'great power' contest between the U.S. and China. The New York Times. http://www.nytimes.com/2021/01/24/us/politics/trump-biden-pentagon-space-missiles-satellite.html

Buss, D. (2005). The murderer next door, New York, NY: Penguin Books.

Callaway, E. (2008, October 13). Loving bonobos have a carnivorous dark side. New Scientist. http://www.newscientist.com/article/dn14926-loving-bonobos-have-a-carnivorous-dark-side

Callaway, E. (2008, October 13). Loving bonobos have a carnivorous dark side. New Scientist. http://www.newscientist.com/article/dn14926-loving-bonobos-have-a-carnivorous-dark-side

Callaway, E. (2015, April 21). Oldest stone tools raise questions about their creators. Nature, 520(7548). http://www.nature.com/news/oldest-stone-tools-raise-questions-about-their-creators-1.17369

Callaway, E. (2017, June 7). Oldest homo sapiens fossil claim rewrites our species' history. Nature. http://www.nature.com/news/oldest-homo-sapiens-fossil-claim-rewrites-our-species-history-1.22114

Caminetti, C. (2020, October). Two women win the Nobel prize in chemistry. Inside Hook http://www.insidehook.com/daily_brief/science/two-women-win-nobel-prize-chemistry

Cherry, K. (2019, June 11). How many neurons are in the brain? International Journal of Tryptophan Research. http://www.ncbi.nim.nih.gov/pmc/articles/PMC5417583/

ChildCare.gov. (2020). Get help paying for childcare. Administration for Children & Families, Office of Child Care. http://www.childcare.gov/consumer-education/ get-help-paying-for-child-care

Choi, C.Q. (2017). Fossil reveals what last common ancestor looked like. Scientific American. http://www.scientificamerican.com/article/fossil-reveals-what-last-common-ancestor-of-humans-and-apes-looked-liked

Cobbresearchlab.com. (2015, Dec. 24). Average cranium/ brain size of Homo neanderthalensis vs. Homo sapiens. Cobbresearchlab. http://www.cobbresearchlab.com/issue-2-1/2015/12/24/average-cranium-brain-size-of-homo-neanderthalensis-vs-homo-sapiens

Cohn, J., Solomon, N. (1994). 30-year anniversary: Tonkin gulf lie launched Vietnam war. Fair. http://www.fair.org/media-beat-column/30-year-anniversary-tonkin-gulf-lie-launched-vietnam-war

Columbia.edu. (2020). Male vs. Female: The brain differences. Columbia. http://www.columbia.edu/itc/anthropology/v1007/jakabovics/mfintro.html

Couppis, M., C, Kennedy. (2008, February 26). Dopamine and the positively reinforcing properties of aggression. Thesis by Maria Couppis at Vanderbilt University. http://www.ir.vanderbilt.edu/handle/1803/10573

Crawford, N. (2018). United States budgetary cost of the post-9/11 wars through FY2019: $5.9 trillion spent and obligated. Watson Institute, Brown University. http://www.watson.brown.edu/costsofwar/files/cow/imce/papers/2018/Crawford_Costs%20of%20War%20Estimates%20Through%20FY2019.pdf

Crawford, N. (2019, November 13). 20 years of war. A costs of war research series. Watson Institute, Brown University. http://www.bu.edu/pardee/research/20-years-of-war-a-costs-of-war-research-series

Dart, R. (1925). Australopithecus africanus: The man-ape of South Africa. Nature. 115, 195-199. http://www.nature.com/articles/115195a0

De Dreu, C. K. W., Greer, L. L., (2011, January 25). Oxytocin promotes human ethnocentrism. Proceedings of the National Academy of Sciences (PNAS), 108(4). http://www.pnas.org/content/108/4/1262

Deneris, E. S., Hendricks, T. J. (2003, January). Pet-1 ETS gene plays a critical role in 5-HT neuron development and is required for normal anxiety-like and aggressive behavior. Neuron. 37(2). http://www.ncbi.nlm.nih.gov/pubmed/12546819

Derleth, J. (2020, September-October). Russian new generation warfare. Military Review. http://www.armyupress.army.mil/Portals/7/militaryreview/Archives/English/SO-20/Derleth-New-Generation-War.pdf

Diamond, J. (2005). Collapse: How societies choose to fail or succeed. New York, NY: Penguin Books.

Diamond, J. (2018, April 22). Origin story. The New York Times Book Review.

Editorial Board. (2020, August 6). The world can still be destroyed in a flash. The New York Times. http://www.nytimes.com/2020/08/06/opinion/hiroshima-anniversary-nuclear-weapons.html

Editors, (2021, June 25). The future of drone warfare. The Week.

Editors, Archaeology. (2014, July 14). The skeletons of Jebel Sahaba. Archaeology. https://www.archaeology.org/news/2305-140714-egypt-conflict-cemetery

Editors, Encyclopedia Britannica. (2018). Mount Toba. Britannica. http://www.britannica.com/place/Mount-Toba/additional-info#history

Editors, history.com. The Bronze Age. History. http://www.history.com/topics/pre-history/bronze-age

Elanger, S. (2019, August 8). Are we headed for another expensive nuclear arms race? Could be. The New York Times. http://www.nytimes.com/2019/08/08/world/europe/arms-race-russiachina.html

Eliot, L. (2021). Brain development and physical aggression. University of Chicago Press Journals. http://www.doi.org/10.1086/711705

Ember, S., Lerer, L. (2020, October 5). Kamala Harris's doubleheader: A debate and hearing with sky-high stakes. The New York Times. http://www.nytimes.com/2020/10/05/us/politics/kamala-harris-debate.html

Encyclopedia Britannica. http://www.britannica.com/list/8-deadliest
-wars-of-the-21st-century

Ferguson, R. B. (2003, July/August). The birth of war. Natural History. http://www.academia.edu/3112521/The_Birth_of_War

Fine, C. (2017). Testosterone rex: Myths of sex, science, and society.
New York, NY: W. W. Norton & Company.

French, D. (2019, January 7). Grown men are the solution, not the
problem. National Review. https://www.nationalreview.com
/2019/01/psychologists-criticize-traditional-masculinity

Fry, D. (2007). Beyond war: The human potential for peace. New
York, NY: Oxford University Press.

Fukuyama, F. (1998, September/October). Women and the evolution of world politics. Foreign Affairs. http://www.foreignaffairs
.com/authors/francis-fukuyama

Ganna, A. (2019, August 30). Large-scale GWAS reveals insights
into the genetic architecture of same-sex sexual behavior. Science,
365(645).

Garber, M. (2013, Sept. 26). The man who saved the world by doing absolutely nothing. The Atlantic. http://www.theatlantic.com
/technology/archive/2013/09/the-man-who-saved-the-world-by-do-
ing-absolutely-nothing/280050/

Gates, R. (2020). Exercise of power: American failures, successes,
and a new path forward in the post-cold war world. New York NY:
Alfred A. Knopf/Penguin Random House.

Gilby, I. (2014, Sept. 18). Study finds lethal aggression is natural in chimpanzees. Arizona State University News. http://www.asunow .asu.edu/ content/study-finds-lethal-aggression-natural-chimpanzees

Glausiusz, J. (2020). Would the world be more peaceful if there were more women leaders? Aeon.org. https://aeon.co/users /josie-glausiusz

Goldman, B. (2017). Two minds. The cognitive differences between men and women. Stanford Medicine. http://www.stanmed.stanford.edu/2017spring/how-mens-and-womens-brains-are-different.html

Goldstein, J. (2012). Female Combatants. From The Encyclopedia of War. Hoboken, NJ: Blackwell Publishing Ltd. http://www .warandgender.com/goldstein%20female%20combatants.pdf

Goodall, J. (1986). The chimpanzees of Gombe: Patterns of behavior. New York, NY: Belknap Press of Harvard University Press.

Goodall, J. (2000). Through a window: My thirty years with the chimpanzees of Gombe. Boston, NY: Mariner/Houghton Mifflin Company.

Gun Violence Archive. (2020). Past summary ledgers. Gun Violence Archive. http://www.gunviolencearchive.org/past-tolls

Guttmacher Institute. (2018, May). Contraceptive use in the United States by demographics. Guttmacher Institute. http://www .guttmacher.org/fact-sheet/contraceptive-use-united-states

Hamer, D. (1993). A linkage between DNA markers on the X chromosome and male sexual orientation. Science, 16;261 (5119), 321-7. http://www.ncbi.nlm.nih.gov/pubmed/8332896

Handelsman, D., Hirschberg, A., Bermon, S. (2018, October). Circulating testosterone as the hormonal basis of sex difference athletic performance. Endocrine Reviews. http://www.ncbi.nlm.nih.gov/pmc/articles/PMC6391653/

Handwerk, B. (2016, January 20). An ancient, brutal massacre may be the earliest evidence of war. Smithsonian Mag. http://www.smithsonianmag.com/science-nature/ ancient-brutal-massacre-may-be-earliest-evidence-war-180957884/

Harmand, S. (2015). 3.3-million-year-old stone tools from Lomekwi 3, West Turkana, Kenya. Nature. 521, 310-315. http://www.nature.com/articles/nature14464

Harari, Y. (2018). Sapiens. A Brief History of Humankind. New York, NY: Harper Perennial

Hart, D., Sussman, R. (2009). Man the hunted: Primates, predators, and human evolution. Boulder, CO: Westview Press.

Hatemi, P and McDermott, R. (2020, December 4). Revenge is a dish best served nuclear. U.S. deterrence depends on it. Bulletin of the Atomic Scientists. http://www.thebulletin.org/2020/12/revenge-is-a-dish-best-served-nuclear-us-deterrence-depends-on-it

Headlines Hopkins. (1995, November). Scientists discover a genetic basis for aggressive behavior in male mice. JH. http://www.pages.jh.edu/news_info/news/home95/nov95/mice.html

Healy M. (2015 October 8). Scientists find DNA differences between gay men and their straight twin brothers. Los Angeles Times. http://www.latimes.com/science/sciencenow/la-sci-sn-genetic-homosexuality

Hedges, C. (2003). What every person should know about war. New York, NY: Free Press/Simon & Schuster.

Hedges, C. (2003). What every person should know about war. New York, NY: Free Press/Simon & Schuster.

Hedges, C. (2003). What every person should know about war. The New York Times. http://www.what-every-person-should-know-about-war.html

Hedges, C. (2003). What everybody should know about war. New York, NY: Free Press/Simon & Schuster.

History.com editors. (2020). Civil war. History. http://www.history.com/topics/american-civil-war/American-civil-war-history

History.com editors. (2020). Korean war. History.com. http://www.history.com/topics/korea/korean-war

Hooven, C. (2021). T: The Story of Testosterone, the Hormone That Dominates and Divides Us. New York, NY: Henry Holt & Company

Hublin, J. J. (2017). New fossils from Jebel Irhoud, Morocco, and the pan-African origin of Homo sapiens. Nature, 546, 289–292. http://www.ncbi.nlm.nih.gov/pubmed/28593953

Human cost of post – 9/11 wars: Direct war deaths in major war zones. Watson Institute, Brown University. https://watson.brown .edu/costsofwar/figures/2019/direct-war-death-toll-2001-801000

Ireland, C. (2008, April 3). Eating meat led to smaller stomachs, bigger brains. Harvard news. https://news.harvard.edu/gazette/story /2008/04/eating-meat-led-to-smaller-stomachs-bigger-brains

Isaacson, W. (2007). Einstein: His life and universe, New York, NY: Simon & Schuster.

Jemison, M. (2014, July). Human evolution rewritten: We owe our existence to our ancestor's flexible response to climate change. Smithsonian Insider. https://insider.si.edu/2014/07/human -evolution-rewritten-flexible-response-climate-change

Keating, K. (2016, June 9). Family leave in the U.S. and Europe: A comparison. Huff Post. http://www.huffpost.com/entry/family -leave-in-the-us-an_b_7543298

Keegan, J. (1994). A history of warfare. New York, NY: Vintage Books.

Kellerman, A., Mercy, J. (1992). Men, women, and murder: gender-specific differences in rates of fatal violence and victimization. NCBI/PubMed. http://www.ncbi.nlm.nih.gov/pubmed/1635092

Kiger, P. (2019, June 17). Key moments in the Cuban missile crisis. History.com. http://www.history.com/news/cuban-missile-crisis -timeline-jfk-khrushchev

King, J., Johnson, D., VanVugt, M. (2009, October 13). The origins and evolution of leadership. Current Biology. http://www .sciencedirect.com/science/article/pii/S0960982209014122

Klinghoffer, D. (2018, July 31). Geneticist: on human-chimp genome similarity, there are "predictions," not "established fact. Evolution News. http://www.evolutionnews.org/2018/07/geneticist-on -human-chimp-genome-similarity-there-are-predictions-not -established-fact

Klyuchnikova, M.A., Voznesenskaya, V.V. (2011). Genetic regulation of intermale aggression in the house mouse. Doklady Biological Sciences, 436, 26-28. http://www.ncbi.nlm.nih.gov/pubmed /21374007

Kouwenhoven, A. (1997, May/June). World's oldest spears. Archaeology New Briefs. 50(3). http://www.archive.archaeology.org/9705 /newsbriefs/spears.html

Krause, J. (2007, Nov. 6). The derived FOXP2 variant of modern humans was shared with neandertals. Current Biology. doi.org/10 .1016/j.cub.2007.10.008

Kravitz, E. Kravitz Lab, (2003). page 1, http://www.hms.harvard.edu /bss/neuro/kravitz/

Labash, M. (2019, October 24). Not your father's masculinity. The New York Times.

Le Marquand, D. (2008). Biochemical factors in aggression and violence. Encyclopedia of Violence, Peace & Conflict 2nd ed.

LeBlanc, S. (2003). Constant battles: The myth of the peaceful, noble savage. New York, NY: St. Martin's Press.

LeVay S. (1991). A difference in hypothalamic structure between heterosexual and homosexual men. Science, 253(5023), 1034. https://science.sciencemag.org/content/253/5023/1034

Levine, P., McKnight, R. (2020). Three million more guns: The Spring 2020 spike in firearm sales. Brookings. http://www .brookings.edu/blog/up-front/2020/07/13/ three-million-more-guns -the-spring-2020-spike-in-firearm-sales

Lewis, J., Harmand, S. (2015, May 20). Our stone tool discovery pushes back he archaeological record by 700,000 years. The Conversation. http://www.theconversation.com/our-stone-tool -discovery-pushes-back-the-archaeological-record-by-700-000 -years-42103

Longman, J. (2019, July 30). Caster Semenya barred from 800 meters at world championships. The New York Times.

Longrich, N. (2019, November 22). Nine species of human once walked earth. Now there's just one. Did we kill the rest? Science Alert. http://www.sciencealert.com/did-homo-sapiens-kill-off-all -the-other-humans

Lorenz, K. (1996). On aggression. New York, NY: Harcourt, Brace & World, Inc. New York, NY: Houghton Mifflin Company.

Lovinger, P. (2019, June 30). Presidential war powers and Bill Clinton's battles. History News Network. http://www.historynewsnetwork.org/article/172398

Mann, C. (2019, April 18). Congressional research service. FAS. http://www.fas.org/sgp/crs/natsec/IF11182.pdf

Mansfield, M. (2019). Startup statistics – The numbers you need to know. Small Biz Trends. http://www.smallbiztrends.com/2019/03 /startup-statistics-small-business.html

Marean, C. and Brown, L. (2012, November 22). Early projectile weapons. Nature, 491(7425), 22

Markoff, J. (2015, October 22). Sorry, Einstein, but 'spooky action' seems real. The New York Times

Marshall, M. (2009, July 7). Timeline: Weapons technology. New Scientist. http://www.newscientist.com/ article/dn17423-timeline -weapons-technology

Marshall, M. (2015, July 28). Chimpanzees over-hunt monkey prey almost to extinction. BBC. http://www.bbc.com/earth/story /20150728-chimps-nearly-wiped-out-monkeys

Mashal, M. (2020, May 26). How the Taliban outlasted a super-power in Afghanistan. The New York Times. http://www.nytimes .com/2020/05/26/world/asia/taliban-afghanistan-war.html

McDermott, R. (2008, November 15). Born to fight, evolved for peace. (main article by Holmes, B). New Scientist. http://www .sciencedirect.com/science/article/abs/pii/S0262407908628560

McDermott, R., Tingley, D., (2008, December). Monoamine oxidase a gene (MAOA) predicts behavioral aggression following provocation. Proceedings of the National Academy of Sciences (PNAS). http://www.pnas.org/content/106/7/2118

Mecklin, J. (2020, January). Closer than ever: It is 100 seconds to midnight. Bulletin of the Atomic Scientists. http://www.thebulletin .org/doomsday-clock/current-time/

Meisenberg, G. In God's Image? The Natural History of Intelligence and Ethics. Sussex, England: Book Guild Publishing

Milton, K. and McBroom, P. (1999, June 14). Meat-eating was essential for human evolution, says UC Berkeley anthropologist specializing diet. http://www.berkeley.edu/news/media/releases /99legacy/6-14-1999a.html

Mitani, J., Watts, D., Amsler, S. (2010, June). Lethal inter-group aggression leads to territorial expansion in wild chimpanzees. Current Biology. 20(12).doi:10.1016/j.cub.2010.04.021

Morelle, R. (2015, May 20). Oldest stone tools pre-date earliest humans. BBC. http://www.bbc.com/news/science-environment -32804177

Morgan, D. (2014, Sept 17). Nature of war: Chimps inherently violent; Study disproves theory that 'chimpanzee wars' are sparked by human influence. Science Daily. http://www.sciencedaily.com /releases/2014/09/140917131816.html

Mulrine, A. (2012, October 16). Cuban missile crisis: The 3 most surprising things you didn't know. The Christian Science Monitor. http://www.csmonitor.com/USA/ Politics/DC-Decoder/2012

/1016/Cuban-Missile-Crisis-the-3-most-surprising-things-you-didn-t
-know/The-Cuban-Missile-Crisis-almost-caused-a-US-military-coup

National Association of Women Business Owners. (2017). Women
business owner statistics. National Association of Women Business
Owners. http://www.nawbo.org/resources/women-business-owner
-statistics

National Human Genome Research Institute. (2010). Why Mouse
Matters. Genome. https://www.genome.gov/10001345
/importance-of-mouse-genome

Nelson, R. J., Snyder, G. Demas, (1999). Elimination of aggressive
behavior in male mice lacking endothelial nitric oxide synthase.
The Journal of Neuroscience, 19:RC30:1-5. http://www.ncbi.nlm
.nih.gov/pubmed/10493775

Nelson, R.J., Chiavegatto. (2001, December). Molecular basis of aggression. Trends in neurosciences, 24(12): 713-9s. http://www.ncbi
.nlm.nih.gov/pubmed/11718876

Ngun, T. C., Vilain, E. (2014). The biological basis of human sexual
orientation: Is there a role for epigenetics? Advances in Genetics,
86, 167-84. http://www.PubMed.gov

North, N. (2017, October 6). Birth control saves lives. Trump just
made it harder to get. Vox. http://www.vox.com/identities/2017
/10/6/16243980/ birth-control-mandate-trump

NPR staff. (2011, January 17). Ike's warning of military expansion,
50 years later http://www.npr.org/2011/01/ 17/132942244
/ikes-warning-of-military-expansion-50-years-later

O'Connor, D. B., Neave, N. (2009). Testosterone and male behaviours. The Psychologist, 22, 28-31. http://www.researchgate.net/publication/ß258689495_Testosterone_and_male_behaviours

Otterbein, K. (2004). How war began. College Station. Texas, TX: A&M University, University Press.

Patton, G. (1944, June 5). As compiled by Charles M. Province. Patton HQ. http://www.pattonhq.com/speech.html

Pbs.org. Evolution: Origins of humankind. PBS. http://www.pbs.org/wgbh/evolution/humans/

Phelps, S. M., Wedow, R. (2019, August 29). What genetics is teaching us about sexuality. The New York Times. http://www.nytimes.com/2019/08/29/ opinion/ genetics-sexual-orientation-study.html

Plomin, R. (2018). Blueprint: How DNA makes us who we are. UK, USA: Allen Lane (Penguin Random House).

Pobiner, B. (2016). Meat-eating among the earliest humans. American Scientist. 104(2), 110. http://www.americanscientist.org/article/meat-eating-among-the-earliest-humans

Pontzer, H. (2019, January). Evolved to exercise. Scientific American. 23-29. http://www.scientificamerican.com/article/humans-evolved-to-exercise

Potts, M., Hayden, T. (2008). Sex and war. Dallas, TX: Ben Bella Books, Inc.

Powell, B. (2004). U.N. weapons inspector Hans Blix faults Bush administration for lack of 'critical thinking' in Iraq. Berkeley. http://www.berkeley.edu/news/media/releases/2004/03/18_blix .shtml

Pruetz, J. (2017). ISU anthropologist's study is first to report chimps hunting with tools. Iowa State University News. http://www .iastate.edu/news/2007/feb/ chimpstools.shtml

Putt, S. (2017, May 8). The functional brain networks that underlie Early Stone Age tool manufacture. Nature. Human Behavior. 1(0102). http://www.nature.com/articles/s41562-017-0102

Ray, M. (2020). 8 deadliest wars of the 21st Century. Encyclopedia Britannica. http://www.britannica.com/list/8-deadliest-wars-of-the -21st-century

Reardon, S. (2019 August 29). Massive study finds no single genetic cause of same-sex sexual behavior. Scientific American. http://www .scientificamerican.com/ article/massive-study-finds-no-single -genetic-cause-of- same-sex-sexual-behavior/

Roach, J. (2007, July). Chimps use "spears" to hunt mammals, study says. National Geographic. http://www.nationalgeographic.com/sci- ence/2007/02/chimps-use- spears-to-hunt-mammals-study-says /Smithsonian

Roach, N. (2013, June 26). Elastic energy storage in the shoulder and the evolution of high-speed throwing in Homo. Nature. http://www.nature.com/articles/nature12267

Rutgers Graduate School of Education. (2019, January 31). Research shows high-quality pre-k can pay off. Now let's deliver it. National Institute for Early Education Research. http://www.nieer.org/2019/01/31/research- shows-high-quality-pre-k-pays-off-now-lets-deliver-it

Sale, K. (2006). After Eden: The evolution of human domination. Durham, NC: Duke University Press.

Sapolsky, R. (1997). The trouble with testosterone. New York, NY: Scribner.

Sapolsky, R. (2018) Behave: The Biology of Humans at Our Best and Worst: New York, NY: Penguin Books.

Schriock, S., Reynolds, C. (2021). Run to win: Lessons in leadership for women changing the world. New York, NY: Dutton, imprint of Penguin Random House.

Schuster, C. (2008). Case closed: The gulf of Tonkin incident. http://www.historynet.com/case-closed-the- gulf-of-tonkin -incident.html

ScienceDaily.com. (2008, January 25). Aggression as rewarding as sex, food and drugs, new research shows. Vanderbilt University. ScienceDaily. http://www.ts-si.org/soc-&-psych/2873

Sciencedaily.com. (2014, July 3). Bone marrow fat tissue secretes hormone that helps body stay healthy. Science Daily. http://www .sciencedaily.com/releases/2014/07/140703125216.html

Sharkey, N. (2020, February). Fully autonomous weapons pose unique dangers to humankind. Scientific American.

Smith, D. (2007). The most dangerous animal; human nature and the origins of war. New York, NY: St. Martin's Press.

Smith, R. (2019, June 19). Hypersonic missiles are unstoppable and they're starting a new global arms race. The New York Times Magazine.

Stanford, C. (1998 August-October). The social behavior of chimpanzees and bonobos. Current Anthropology. 39(4). http://www.psycnet.apa.org/record/1999-13940-001

Stanford, C. (1999). The hunting apes: Meat eating and the origins of human behavior. Princeton, NJ: Princeton University Press.

Storr, A (1968). Human Aggression. New York, NY: Atheneum

Strauss, M. (2015). 12 theories of how we became human, and why they're all wrong. National Geographic. http://www.nationalgeographic.com /news/2015/09/150911-how-we-became-human-theories-evolution-science

Taub, A. (2020, August 13). Why are women-led nations doing better with Covid-19? The New York Times. http://www.nytimes.com/2020/05/15/world/coronavirus-women-leaders.html

The Beijing Platform for Action Turns 20. (2020). Women of achievement. Beijing 20. http://www.beijing20.unwomen.org/en/voices-and-profiles/women-of-achievement

Thomasello, M. (2019). Becoming human: A theory of ontogeny. New York, NY: Harvard University Press.

Thompson, J. (2019, February). Origins of the human predatory pattern: The transition to large-animal exploitation by early hominins. Current Anthropology. 60(1), 1- 23. http://www.doi.org/10.1086/701477

Thompson, J. (2019). Origins of the human predatory pattern: The transition to large-animal exploitation by Early Hominins. University of Chicago Press Journals. Current Anthropology. 60(1). http://www.journals.uchicago.edu/doi/10.1086/701477

Towle, I. (2017 July 14). Chipped teeth suggests [sic] Homo naledi had a unique diet. The Conversation. http://www.theconversation.com/chipped-teeth-suggests-homo-naledi-had-a-unique-diet-80714

Tullis, P. (2019, December). GPS down: Hacking the system we all rely on is not difficult, and the U.S. has no defense in place. Scientific American.

U. S. Department of Defense. (2020, August 3). Casualty status. U. S. Department of Defense. http://www.defense.gov/casualty.pdf

University of Chicago Medical Center/ScienceDaily. (2010, March 10). Life is shorter for men, but sexually active life expectancy is longer. Science Daily. http://www.sciencedaily.com/releases/2010/

Uphadhayya N., Guragain S. (2014). Comparison of cognitive functions male and female medical students: A pilot study. Journal of Clinical & Diagnostic Research. http://www.ncbi.nlm.nih.gov/pubmed/25120970

W. Rice., U. Friberg., S. Gavrilets. (2012). Homosexuality as a consequence of epigenetically canalized sexual development. The Quarterly Review of Biology, 87(4), 343-368.

Walsh, J. (2011, August 22). Man entered the kitchen 1.9 million years ago. Proceedings of the National Academy of Sciences. http://www.livescience.com

Ward, T. (2017, May 27). Watch: Do we really share 99% of our DNA with chimps? Futurism. http://www.futurism.com/watch -do-we-really-share-99-our-dna-chimps.

Wars in the World. (2020, January 21). List of ongoing conflicts. Wars in the World. http://www.warsintheworld.com/?page =static1258254223

Wars in the World. (2020, January 21). List of ongoing conflicts. http://www.warsintheworld.com/?page=static1258254223

WashingtonTimes.com. Game Changer: America's most advanced weapons. Washington Times. http://www.washingtontimes .com/multimedia/collection/game-changer-americas-most -advanced-weapons

Watson Institute, Brown University. (2019, November). Human cost of post–9/11 wars: Direct war deaths in major war zones. Watson Institute, Brown University. https://watson.brown.edu/costsofwar /figures/2019/direct-war-death-toll-2001-801000

Watson Institute, Brown University. (2019). Costs of war project. Watson Institute, Brown University. http://www.watson.brown .edu/costsofwar

Watts, D. (2006). Lethal intergroup aggression by chimpanzees in Kibale National Park, Uganda. American Journal of Primatology. http://www.onlinelibrary.wiley.com/doi/abs/10.1002/ajp.20214

Wersinger, S.R., Ginns, E. I. (2002, November). A genetic basis for aggression and anger. Molecular Psychiatry. http://www.home95 /nov95/mice.htmljhu.edu/news_info/news/

Whittaker, D. (2012, October 1). Evolution 101: Natural selection. Beacon Center. http://www.3.beacon-center.org/page /37

Wilkins J. (2017, February). Middle Pleistocene lithic raw material foraging strategies at Kathu Pan 1, Northern Cape, South Africa. Journal of Archaeological Science: Reports. http://www .sciencedirect.com/science/article/abs/pii/S2352409X16304308

Williams, A.C., Hill, L. J. (2017). Meat and nicotinamide: A causal role in human evolution, history, and demographics. NCI. http://www.ncbi.nlm.nih.gov/pubmed/28579800

Wilson, E. O. (1975). Sociobiology: The new synthesis. Cambridge, MA: Harvard University Press.

Wilson, V., Wilson, J. (2013). How the Bush administration sold the war–and we bought it. The Guardian. http://www.theguardian .com/commentisfree/2013/feb/27/bush-administration-sold -iraq-war

Wolfe, T. (2006). The Right Stuff. New York, NY: Bantam Books.

Wong, K. (2012, November 15). Human ancestors made deadly stone-tipped spears 500,000 years ago. Scientific American.

http://www.blogs.scientificamerican.com/observations/human
-ancestors-made-deadly-stone-tipped-spears-500000-years-ago

Wong, K. (2014, April). Hunting was a driving force in human evolution. Scientific American. 310(4). http://www.scientificamerican
.com/article/hunting-was-a-driving-force-in-human-evolution

Wong, K. (2015). Archaeologists take wrong turn, find world's oldest stone tools [update]. Scientific American. http://www
.scientificamerican.com/observations/archaeologists-take-wrong-
turn-find-world-s-oldest-stone-tools-update/

Wrangham R. Peterson, D. (1996). Demonic males: Apes and the
origins of human violence. New York, NY: Houghton Mifflin Company.

Wrangham, R. (2009). Catching fire: How cooking made us human.
New York, NY: Basic Books.

Wrangham, R. (2019). The goodness paradox: The strange relationship between virtue and violence in human evolution. New York,
NY: Pantheon Books.

Zorich, Z. (2013, March/April). The first spears. Archaeology.
http://www.archaeology.org/issues/81-1303/trenches/523
-south-africa-earliest-spears

# Index

Printed in the USA
CPSIA information can be obtained
at www.ICGtesting.com
CBHW040040240724
12088CB00018B/627